Methods in Molecular Biology

Series Editor
John M. Walker
School of Life and Medical Sciences
University of Hertfordshire
Hatfield, Hertfordshire, UK

For further volumes:
http://www.springer.com/series/7651

For over 35 years, biological scientists have come to rely on the research protocols and methodologies in the critically acclaimed *Methods in Molecular Biology* series. The series was the first to introduce the step-by-step protocols approach that has become the standard in all biomedical protocol publishing. Each protocol is provided in readily-reproducible step-by-step fashion, opening with an introductory overview, a list of the materials and reagents needed to complete the experiment, and followed by a detailed procedure that is supported with a helpful notes section offering tips and tricks of the trade as well as troubleshooting advice. These hallmark features were introduced by series editor Dr. John Walker and constitute the key ingredient in each and every volume of the *Methods in Molecular Biology* series. Tested and trusted, comprehensive and reliable, all protocols from the series are indexed in PubMed.

Metabolic Pathway Engineering

Edited by

Michael E. Himmel and Yannick J. Bomble

National Renewable Energy Laboratory, Golden, CO, USA

Humana Press

Editors
Michael E. Himmel
National Renewable Energy Laboratory
Golden, CO, USA

Yannick J. Bomble
National Renewable Energy Laboratory
Golden, CO, USA

ISSN 1064-3745 ISSN 1940-6029 (electronic)
Methods in Molecular Biology
ISBN 978-1-0716-0194-5 ISBN 978-1-0716-0195-2 (eBook)
https://doi.org/10.1007/978-1-0716-0195-2

This Humana imprint is published by the registered company Springer Science+Business Media, LLC, part of Springer Nature.
The registered company address is: 1 New York Plaza, New York, NY 10004, U.S.A.

Preface

This book illustrates experimental and computational methodologies used to achieve cost-effective biological processes for the production of fuels and biochemicals. There are multiple approaches to increasing yield, titers, and productivity in a robust host. Here, we include the most recent and cutting-edge aspects of pathway engineering, flux analysis, and metabolic enzyme engineering. Each chapter highlights the complexity/challenges of the problem and the methods used to solve this problem or changes needed in current methods. We expect that this book will benefit not only scientists working on more fundamental aspects of this endeavor but also those in the biochemical industry working on strain engineering for robust industrial processes.

Golden, CO, USA

Michael E. Himmel
Yannick J. Bomble

Contents

Contributors

PAUL E. ABRAHAM • *Chemical Sciences Division, Oak Ridge National Lab, Oak Ridge, TN, USA*

MICHAEL W. W. ADAMS • *Department of Biochemistry and Molecular Biology, University of Georgia, Athens, GA, USA*

MARKUS ALAHUHTA • *National Renewable Energy Laboratory, Golden, CO, USA*

YANNICK J. BOMBLE • *National Renewable Energy Laboratory, Golden, CO, USA*

STEVEN D. BROWN • *Lanza Tech, Skokie, IL, USA*

YAT-CHEN CHOU • *National Renewable Energy Laboratory, Golden, CO, USA*

DAEHWAN CHUNG • *National Renewable Energy Laboratory, Golden, CO, USA*

SHARLEE CLIMER • *University of Missouri-St. Louis, St. Louis, MO, USA*

JINGXUAN CUI • *Thayer School of Engineering at Dartmouth, Hanover, NH, USA; Center for Bioenergy Innovation, Oak Ridge, TN, USA*

ASELA DASSANAYAKE • *Lanza Tech, Skokie, IL, USA*

BRYON S. DONOHOE • *Biosciences Center, National Renewable Energy Laboratory, Golden, CO, USA*

CARRIE ECKERT • *National Renewable Energy Laboratory, Golden, CO, USA; Renewable and Sustainable Energy Institute, University of Colorado, Boulder, CO, USA*

EMILY FREED • *National Renewable Energy Laboratory, Golden, CO, USA; Renewable and Sustainable Energy Institute, University of Colorado, Boulder, CO, USA*

ALIDA T. GERRITSEN • *National Renewable Energy Laboratory, Golden, CO, USA*

MICHAEL T. GUARNIERI • *National Renewable Energy Laboratory, Golden, CO, USA*

CALVIN A. HENARD • *National Renewable Energy Laboratory, Golden, CO, USA*

ROBERT L. HETTICH • *Chemical Sciences Division, Oak Ridge National Lab, Oak Ridge, TN, USA*

MICHAEL E. HIMMEL • *National Renewable Energy Laboratory, Golden, CO, USA*

SHUEN HON • *Thayer School of Engineering at Dartmouth, Hanover, NH, USA; Center for Bioenergy Innovation, Oak Ridge, TN, USA*

DANIEL JACOBSON • *Biosciences Division, Oak Ridge National Laboratory, Oak Ridge, TN, USA; The Bredesen Center for Interdisciplinary Research and Graduate Education, University of Knoxville Tennessee, Knoxville, TN, USA*

HUAIGUANG JIANG • *National Renewable Energy Laboratory, Golden, CO, USA*

PIET JONES • *Biosciences Division, Oak Ridge National Laboratory, Oak Ridge, TN, USA; The Bredesen Center for Interdisciplinary Research and Graduate Education, University of Knoxville Tennessee, Knoxville, TN, USA*

ERIC P. KNOSHAUG • *National Renewable Energy Laboratory, Golden, CO, USA*

CHIEN-YUAN LIN • *Biosciences Center, National Renewable Energy Laboratory, Golden, CO, USA*

VLADIMIR V. LUNIN • *National Renewable Energy Laboratory, Golden, CO, USA*

LEE R. LYND • *Thayer School of Engineering at Dartmouth, Hanover, NH, USA; Center for Bioenergy Innovation, Oak Ridge, TN, USA*

PINCHING MANESS • *National Renewable Energy Laboratory, Golden, CO, USA*

SHILPA NAGARAJU • *Lanza Tech, Skokie, IL, USA*

DANIEL G. OLSON • *Thayer School of Engineering at Dartmouth, Hanover, NH, USA; Center for Bioenergy Innovation, Oak Ridge, TN, USA*
YUNQIAO PU • *Biosciences Division, Oak Ridge National Laboratory, Oak Ridge, TN, USA*
MENGYU QIU • *State Key Laboratory of Biocatalysis and Enzyme Engineering, Environmental Microbial Technology Center of Hubei Province, and School of Life Sciences, Hubei University, Wuhan, China*
ARTHUR J. RAGAUSKAS • *Department of Chemical and Biomolecular Engineering, University of Tennessee, Knoxville, TN, USA*
NICHOLAS S. SARAI • *Division of Chemistry and Chemical Engineering, California Institute of Technology, Pasadena, CA, USA*
JEREMY SCHMUTZ • *HudsonAlpha Institute for Biotechnology, Huntsville, AL, USA*
MANESH SHAH • *Biosciences Division, Oak Ridge National Laboratory, Oak Ridge, TN, USA*
ARJUN SINGH • *Denver, CO, USA*
SHARON SMOLINSKI • *National Renewable Energy Laboratory, Golden, CO, USA*
AVINASH SREEDASYAM • *HudsonAlpha Institute for Biotechnology, Huntsville, AL, USA*
PETER C. ST. JOHN • *Biosciences Center, National Renewable Energy Laboratory, Golden, CO, USA*
LIANG TIAN • *Thayer School of Engineering at Dartmouth, Hanover, NH, USA; Center for Bioenergy Innovation, Oak Ridge, TN, USA*
MELVIN P. TUCKER • *National Bioenergy Center, National Renewable Energy Laboratory, Golden, CO, USA*
GERALD TUSKAN • *The Bredesen Center for Interdisciplinary Research and Graduate Education, University of Knoxville Tennessee, Knoxville, TN, USA*
SAGAR UTTURKAR • *Purdue University Center for Cancer Research, West Lafayette, IN, USA*
YU WANG • *State Key Laboratory of Biocatalysis and Enzyme Engineering, Environmental Microbial Technology Center of Hubei Province, and School of Life Sciences, Hubei University, Wuhan, China*
DEBORAH WEIGHILL • *Biosciences Division, Oak Ridge National Laboratory, Oak Ridge, TN, USA; The Bredesen Center for Interdisciplinary Research and Graduate Education, University of Knoxville Tennessee, Knoxville, TN, USA*
HUI WEI • *National Renewable Energy Laboratory, Golden, CO, USA*
AMANDA M. WILLIAMS-RHAESA • *New Materials Institute, University of Georgia, Athens, GA, USA*
WEI XIONG • *National Renewable Energy Laboratory, Golden, CO, USA*
QI XU • *National Renewable Energy Laboratory, Golden, CO, USA*
QING YANG • *State Key Laboratory of Biocatalysis and Enzyme Engineering, Environmental Microbial Technology Center of Hubei Province, and School of Life Sciences, Hubei University, Wuhan, China*
SHIHUI YANG • *State Key Laboratory of Biocatalysis and Enzyme Engineering, Environmental Microbial Technology Center of Hubei Province, and School of Life Sciences, Hubei University, Wuhan, China*

YONGFU YANG • *State Key Laboratory of Biocatalysis and Enzyme Engineering, Environmental Microbial Technology Center of Hubei Province, and School of Life Sciences, Hubei University, Wuhan, China*
CHANG GEUN YOO • *Department of Paper and Bioprocess Engineering, SUNY College of Environmental Science and Forestry, Syracuse, NY, USA*
MIN ZHANG • *National Renewable Energy Laboratory, Golden, CO, USA*
TIANYONG ZHENG • *Thayer School of Engineering at Dartmouth, Hanover, NH, USA; Center for Bioenergy Innovation, Oak Ridge, TN, USA*

Chapter 1

Overview and Future Directions

Yannick J. Bomble and Michael E. Himmel

Abstract

Modern microbial and enzyme engineering and their advancement are increasingly dependent on the marriage of a wide range of sophisticated technologies. For students entering the field of biotechnology, the outlook is indeed daunting. Expertise at levels beyond that of simple familiarity will be needed to conduct competitive research. It goes without saying that to be competitive, all new researchers in this field will need basic preparation in molecular biology, biochemistry, and genetics. Further, experience working with concepts and experimental tools in enzyme biochemistry and kinetics, gene editing, computational metabolic pathway modeling, experimental pathway flux analysis, and computational clustering tools to process complex data sets will be vital for success. We speculate that most biotechnology researchers in early career at this time will build teams of collaborators to address these disparate science fields rather than attempt to become experts in one lab.

Key words Enzyme engineering, Omics, Pathway engineering, Large data sets

1 Introduction

Long extinct are practices of "siloing" education and experimental practice of science skills into fields such as chemistry, biology, physics, mathematics, and mechanical/electrical engineering, etc. Looking to the future of biotechnology, new students must become accomplished in working at the intersections of these classical fields. In the context of this book, for example, studies known as the Omics really focus on the application of experimental tools able to collect large data sets regarding cellular function. Normally, proteomics, transcriptomics, and metabolomics are examples of such approaches. The objective of Omics studies is usually to identify patterns in cellular response to challenges imposed by the researcher (inhibitors, substrates, culture grown conditions, gene editing, etc.), usually following proper baselining of the microbe under study. Mathematical processing of such data is nearly always required to extract the maximum value from such work.

Michael E. Himmel and Yannick J. Bomble (eds.), *Metabolic Pathway Engineering*, Methods in Molecular Biology, vol. 2096, https://doi.org/10.1007/978-1-0716-0195-2_1, © Springer Science+Business Media, LLC, part of Springer Nature 2020

Other examples of Combining powerful tools in science to achieve goals in biotechnology are embedded in work to alter known or introduce new metabolic pathways to industrial microbes. These goals require heavily integrated tools, such as computational pathway modeling and metabolic flux analysis, as well as experimental verification of both the wild type state and the resulting mutant cells. Often, cell-free techniques are used to verify the performance of novel pathways before cloning is attempted. Once the strategies for pathway engineering are decided, genes coding heterologous enzymes can be introduced either in plasmid form or by direct chromosomal integration. As this book addresses, the choice of the enzyme cloned, the assessment of its performance, and the products analyzed are important technologies to be mastered to conduct such work.

When studying the chapters given in this book, the reader is encouraged to consider not only the level of expertise needed to collect data or improved materials (cells, enzymes, products), but also ways that effective teams could be assembled. We see a distinction when considering the application of the techniques exemplified here. Some techniques, such as the Omics, require very complex and costly equipment, coupled with data interpretation conducted by experienced workers using resource-intensive computation algorithms. These methods, however, are distinguished from classical biochemistry in that often they do not deliver "precision results." In other words, if the resulting cluster analysis of a proteomics data set yields 100 or 110 apparent clusters, the interpretation remains the same (*see* by Chap. 15). At the other end of the experimental spectrum, the analysis of the performance of wild type versus mutant enzyme can be very precise. As noted in the chapter by Adams et al., if the researcher determines both the concentration and kinetics of a heterologous enzyme with care, the resulting performance of the enzyme in the pathway of interest can be viewed with confidence and used to tune pathway design.

2 Conclusions

In this book, we bring together important topics that can be summarized as those addressing classical "Omics tools" (Methodologies for Proteomics, Methodologies for Transcriptomics), techniques for studying and engineering metabolic pathways (Bacterial Metabolic Pathways for Fuels, Fungal Metabolic Pathways for Fuels, Algal Metabolic Pathways for Fuels, Kinetic modeling of Metabolic Pathways, and Experimental Flux Analysis), techniques for cloning in key biological systems (Genetics of Unstudied Industrial Microbes, Leaf Protoplast Systems for Switchgrass Transformation, Gene Editing Technologies: Biofuels Pathways), and approaches for improving metabolic enzymes and confirming their performance (Structural Biology of Metabolic Enzymes, Measuring Metabolic Enzyme Performance, Measuring Products), and methods to conduct relatedness modeling of large resulting data (Network Modeling of Complex Data Sets, Connecting Genotype with Phenotype in the Genomics Era).

Acknowledgments

Funding was provided by the BioEnergy Science Center (BESC) and the Center for Bioenergy Innovation (CBI), from the U.S. Department of Energy Bioenergy Research Centers supported by the Office of Biological and Environmental Research in the DOE Office of Science. This work was authored in part by Alliance for Sustainable Energy, LLC, the Manager and Operator of the National Renewable Energy Laboratory for the U.S. Department of Energy (DOE) under Contract No. DE-AC36-08GO28308. The views expressed in the article do not necessarily represent the views of the DOE or the U.S. Government. The U.S. Government retains and the publisher, by accepting the article for publication, acknowledges that the U.S. Government retains a nonexclusive, paid-up, irrevocable, worldwide license to publish or reproduce the published form of this work, or allow others to do so, for U.S. Government purposes.

Chapter 2

Genetics of Unstudied Thermophiles for Industry

Daehwan Chung, Nicholas S. Sarai, Michael E. Himmel, and Yannick J. Bomble

Abstract

Thermophilic organisms hold great potential for industry due to their numerous advantages in biotechnological applications such as higher reaction rate, higher substrate loading, decreased susceptibility to reaction contamination, energy savings in industrial fermentations, and ability to express thermostable proteins that can be utilized in many important industrial processes. Bioprospecting for thermophiles will continue to reveal new enzymatic and metabolic paradigms with industrial applicability. In order to translate these paradigms to production scale, routine methods for microbial genetic engineering are needed, yet remain to be developed in many newly isolated thermophiles. Major challenges and recent developments in the establishment of reliable genetic systems in thermophiles are discussed. Here, we use a hyperthermophilic, cellulolytic bacterium, *Caldicellulosiruptor bescii*, as a case study to demonstrate the development of a genetic system for an industrially useful thermophile, describing in detail methods for transformation, genetic tool utilization, and chromosomal modification using targeted gene deletion and insertion techniques.

Key words Thermophiles, Biofuels, Genetic engineering, *Caldicellulosiruptor bescii*, Restriction-modification system, Thermostable proteins, Electroporation

1 Introduction

Industrial reactions commonly use expensive and toxic transition metal catalysts. In an effort to move toward "green" chemistry and reduce cost, thermostable enzymes have found a niche in that they have the ability to facilitate chemical reactions in aqueous buffers without harsh reaction solvents. Thermophiles and hyperthermophiles, microorganisms that grow optimally above 55 °C and 80 °C, respectively, have attracted intense research interest in recent decades due to their implications for basic science and industry alike [1]. Their appeal to industry stems from the inherent thermostability of their enzymes, especially secreted enzymes that must maintain activity away from the stabilizing environment of the cell, as well as their applicability for whole-cell bioprocessing such as consolidated

Michael E. Himmel and Yannick J. Bomble (eds.), *Metabolic Pathway Engineering*, Methods in Molecular Biology, vol. 2096,
https://doi.org/10.1007/978-1-0716-0195-2_2,

bioprocessing (CBP) that combines high-temperature cellulose degradation and ethanol production [2].

The increasing number of available genome sequences of industrially relevant thermophiles has enabled the discovery of biological meaning in the genomic context. Computational analyses, such as functional genomics and proteomics, have provided opportunities for understanding thermophilic physiology and where it differs from that of mesophiles. Many thermophilic proteins are amenable to characterization in vitro using *Escherichia coli* as an expression host [3]. However, establishment of genetic tools is essential to express a gene of interest in its native thermophilic host and to develop thermophilic biocatalysts to meet the economic demands of industry.

Development of reliable genetic systems in thermophiles has been hampered by several hindrances not encountered in mesophiles. The major challenges in the establishment of genetic tools as well as recent advances in overcoming these obstacles in thermophiles will be discussed. The transfer of foreign DNA into a microbe is the first procedure to begin genetically modifying bacterial cells. Common transformation methods in thermophilic microbes can be classified into five main categories: natural competence [4, 5], conjugation [6], protoplast [7, 8], chemical treatment with $CaCl_2$ and heat shock [9, 10], and electrotransformation [11–14]. For thermophiles, natural competency is the simplest DNA uptake system, but its appearance is rare.

Electrotransformation, the introduction of DNA by introducing pores in the cell membrane with high electric field strengths and exponential decay, is the most commonly used method for newly isolated thermophiles due to its high transformation efficiency compared to other methods and relatively simple procedure [2]. The activity of host restriction-modification systems often places a formidable barrier to the introduction of DNA into cells; thus, identifying restriction-modification systems and overcoming them is prerequisite to developing genetic systems in some microbes. For example, the transformation efficiency of two potent CBP microbes, *Clostridium thermocellum* and *Caldicellulosiruptor bescii*, is strictly governed by using properly methylated foreign DNA during electrotransformation [15, 16].

Construction of replicating shuttle vectors constitutes a very important genetic tool to facilitate the optimization of transformation protocols and to express genes of interest in a physiologically relevant environment. Expression vectors provide a unique opportunity to express and purify natively posttranslationally modified proteins to better understand the nature of thermophilic proteins [3]. The two major challenges impeding thermophilic vector development are the lack of both thermophilic origins of replication and thermostable selective markers. Several thermophilic replication origins are available and have been successfully used in several thermophiles including *Thermoanaerobacterium* spp., *Clostridium* spp., *Thermoanaerobacter* spp., *Caldicellulosiruptor* spp., and *Pyrococcus furiosus* [17–20]. Most shuttle vectors used in mesophiles employ drug resistance for selection, but these selective systems are generally inappropriate in thermophiles mainly due to the thermal instability of both the antibiotics and resistance gene products [2]. Nutritional selection markers based on uracil prototrophic selection are commonly used in thermophilic genetic systems. Recently, thermostable antibiotic-based selection markers have also been developed and are being used to develop autonomously replicating vectors for thermophiles. Several antibiotics have been successfully used in thermophiles, including kanamycin, bleomycin, hygromycin, and simvastatin [2, 3, 21]. Genetic tools for chromosomal modification are also very important to investigate in vivo functions of interesting proteins, and to obtain stable strains to serve as industrial hosts. Currently available gene targeting systems often use homologous recombination along with provision of a selection marker. Gene disruption and insertional mutagenesis using either single crossover and/or double crossover homologous recombination have been successfully performed in several thermophiles using both antibiotic and nutritional selection markers [21].

In this chapter, we focus on genetic manipulation systems for *Caldicellulosiruptor bescii*, which have recently been reliably established with a variety of genetic tools [12, 16, 22]. *C. bescii* is an anaerobic, gram-positive bacterium and the most thermophilic cellulolytic microbe currently published [23, 24]. Growing optimally at ~78 °C, *C. bescii* has been of great interest to industry mainly due to its unique ability to utilize a wide range of substrates including crystalline cellulose, hemi-cellulosic sugars, xylan, and non-pretreated plant biomass such as hardwood poplar, switchgrass, and Napier grasses [23, 25]. Its genome contains a large inventory of thermostable glycoside hydrolases (52 GH), polysaccharide lyases (4 PL), and carbohydrate binding modules (22 CBM); as such, it should be a great reservoir of thermostable Carbohydrate-Active Enzymes [25]. For example, CelA, the most active thermophilic cellulase known, was isolated from *C. bescii* [26]. *C. bescii* was initially classified as *Anaerocellum thermophilum* [27, 28]. Later, phylogenetic analyses reclassified *A. thermophilum* as *C. bescii* resulting from the interest from the BioEnergy Science Center (BESC) in pursuing it as a CBP host due to its extremely cellulolytic phenotype [24]. The development of genetic system in *C. bescii* is a good showcase for the study of the genetics of uncharacterized thermophiles with applicability in industrial processes for the production of valuable bio-derived chemicals, biofuels, and thermostable extracellular enzymes. In the following sections, we provide detailed instructions for transformation of *C. bescii* by electroporation. In addition, we describe materials and methods concerning the use of: defined medium (LOD; low osmolarity defined) for suitable nutrient selection [29], nutritional selection markers [16], shuttle vectors [12], homologous and heterologous expression vectors [30–33], and markerless gene deletion [16, 34, 35].

2 Materials

2.1 Caldicellulosiruptor bescii Strains

The uracil auxotroph *C. bescii* mutant strain, JWCB005 [*ΔpyrFA*], is used for transformation optimization and construction of an *E. coli/C. bescii* shuttle vector [12]. The *C. bescii* mutant strain, JWCB018 [*ΔpyrFA ldh::ISCbe4 Δcbe1*], a uracil auxotroph containing a deletion of the CbeI restriction enzyme, is used for routine transformations to simplify genetic manipulations [16, 36].

2.2 Bacterial Media

Variants of DSMZ 516 media [28], referred to as LOD (low osmolarity defined growth medium) and LOC (low osmolarity complex growth medium), are used [29]. The composition of LOC medium is described in Table 1. LOD and LOC media were prepared from filter sterilized stock solutions; 50× Cbe partial base salt solution [22], 1000× Trace Elements SL-10 (Table 2) [37], and 2000× vitamin mix (Table 3) are prepared as described previously [38]. The chemically defined medium, LOD, was created by removing both yeast extract and casein from LOC media.

2.3 Reagents

1. LOD-AA media—LOD media containing 19 amino acid solution [39], used for culture growth to make competent cells.
2. 20 mM uracil in water—filter sterilized.
3. 10% sucrose—used as a washing buffer during competent cell preparation.
4. 5-Fluoroorotic acid (5-FOA)—used at final concentration of 8 mM.

2.4 Equipment

1. Electroporator (e.g., BioRad Gene Pulser).
2. Incubator with shaker at 75 °C.
3. Heat block or water bath at 50 °C—used during plating.
4. Plastic petri dishes.
5. Sterile toothpicks.

3 Methods

3.1 Shuttle/Expression Vector Plasmid Construction

Plasmid pDCW89 (Fig. 1a) [12], an *E. coli/C. bescii* shuttle vector, was constructed using the replication origin of pBAS2, a native plasmid in *C. bescii* [28]. The homologous protein expression vector, pDCW170 (Fig. 1b), constructed based on pDCW89, was

Table 1
LOC media composition [29]

	LOC	
	g/L	mM
Salts		
NH_4Cl	0.25	4.67
KH_2PO_4	0.0136	0.10
KCl	0.33	4.43
$MgCl_2 \cdot 6H_2O$	0.33	1.62
$CaCl_2 \cdot 6H_2O$	0.14	0.95
$NaHCO_3$	1	11.90
Ions		
NH_4^+		4.67
Na^+		15.90
K^+		4.53
Ca^{2+}		0.95
Mg^{2+}		1.62
Cl^-		14.25
$H_2PO_4^-$		0.10
HCO_3^-		11.90
Organics		
Cellobiose/maltose	5	14.60
Cysteine $HCl \cdot H_2O$	1	5.69
Yeast extract	1	
Casein	2	
Trace elements 1000× [37]	1 mL/L	
Resazurin	0.00025	
Vitamin mix 2000× [38]	0.5 mL/L	
pH[a]	7.2	

[a]Both LOD and LOC are pH adjusted as 7.2 by the addition of 4 M NaOH

successfully utilized for CelA expression in *C. bescii* [30]. Information concerning these vectors is described in Fig. 1. Detailed information for the construction of plasmids has been fully described [12, 30] .

Table 2
Recipe for trace elements SL-10 (1000×) [37]

Components	Amount
HCl (25%; 7.7 M)	10 mL
$FeCl_2 \cdot 4H_2O$	1.5 g
$ZnCl_2$	70 mg
$MnCl_2 \cdot 4H_2O$	100 mg
H_3BO_3	6 mg
$CoCl_2 \cdot 6H_2O$	190 mg
$CuCl_2 \cdot 2H_2O$	2 mg
$NiCl_2 \cdot 6H_2O$	24 mg
$Na_2MoO_4 \cdot 2H_2O$	36 mg
dH_2O (fill up to 1 L)	1000 mL

Table 3
Recipe for vitamin mix (2000×) [38]

Components	Amount
Biotin	40 mg
Folic acid	40 mg
Pyridoxine HCl	200 mg
Thiamine $HCl \cdot 2H_2O$	100 mg
Riboflavin	100 mg
Nicotinic acid	100 mg
D CaPentothenate	100 mg
Vitamin B12	2 mg
p-aminobenzoic acid	100 mg
Lipoic acid	100 mg
dH_2O (fill up to 1 L)	1000 mL

3.2 Transformation Protocol

All transformation experiments were performed under aerobic conditions. For liquid growth cultures, we used anaerobic culture serum bottles or hungate tubes degassed with at least three cycles of vacuum and argon.

3.2.1 Preparing DNA for Electroporation

1. Transformation without M.CbeI methylation. Prepare the DNA using a MiniPrep kit (e.g., Qiagen MiniPrep), elute with 10 mM Tris buffer at pH 8–8.3. Do not use TE or EB buffer.

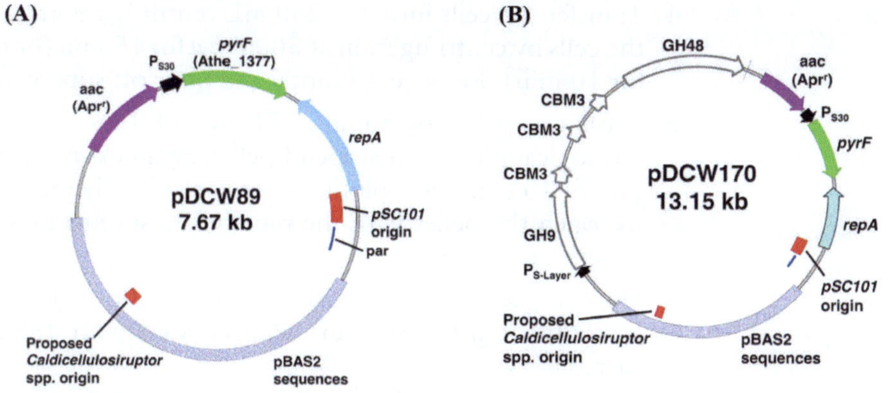

Fig. 1 Plasmid map of (**a**) shuttle vector (pDCW89) and (**b**) CelA expression vector (pDCW170). (**a**) The cross-hatched box corresponds to the pBAS2 plasmid sequences. The apramycin resistant gene cassette (Apr^r); *pSC101* low copy replication origin in *E. coli*; *repA*, a plasmid-encoded gene required for *pSC101* replication; par, partition locus are indicated. The proposed replication origin (115 bp) of pBAS2 is also indicated. A *pyrF* expression cassette along with ribosomal protein S30EA (Cbes2105) promoter is used as a nutritional selection marker. (**b**) An expression cassette, containing the regulatory region of the *C. bescii* S-layer protein, a *C*-terminal 6X His-tag version of CelA (Cbes1867) and a Rho-independent terminator, is inserted between pBAS2 sequences and Apramycin resistance expression cassette in pDCW89

2. In vitro methylation using purified M.CbeI:
   (a) Use DNA isolated from *E. coli* DH5a (*dam*$^+$*dcm*$^+$).
   (b) Incubate 20 mg of DNA with 5 μg of purified M.CbeI [40] in reaction buffer (50 mM Tris–HCl, 50 mM NaCl, 80 mM *S*-adenosylmethionine (SAM), 10 mM Dithiothreitol (DTT) at pH 8.5) in 400 μL total reaction volume for 2 h at 78 °C (*see* **Note 1**).
   (c) Dialyze the methylated DNA sample(s) using a nitrocellulose filter and DI water (*see* **Note 2**).
   (d) The extent of methylation may be evaluated by cleavage using HaeIII and/or CbeI restriction enzymes according to the supplier's instructions.

3.2.2 Preparation of Competent Cells

1. Grow an overnight culture of the required *C. bescii* strain in LOD medium at 75 °C.
2. Inoculate 0.5–1% of the freshly grown JWCB005 (or JWCB018) overnight culture, into 500 mL LOD-AA medium containing 40 μM uracil (*see* **Note 3**). Shaking is optional, although not necessary.
3. When an OD_{680} of 0.1–0.15 is reached in early log phase, take out the growth culture from incubator and cool for 15 min using cool tap water (*see* **Note 4**).

4. Transfer the cells into 2 × 250 mL centrifuge bottles. Harvest the cells by centrifugation at 5000 × *g* for 15 min (or 6000 × *g* for 10 min). Remove promptly and pour off supernatant.
5. Wash cell pellet by adding 250 mL of 10% sucrose. Use a serological pipette to suspend pellet by quick, repeated pipetting of 10% sucrose solution over the pellet, being careful not to touch the pellet with the pipette. Be sure to fully suspend pellet (*see* **Note 5**).
6. Harvest the cells by centrifugation at 5000 × *g* for 15 min (or 6000 × *g* for 10 min). Repeat wash with 250 mL 10% sucrose.
7. Harvest the cells by centrifugation at 5000 × *g* for 15 min (or 6000 × *g* for 10 min). Suspend cells in 50 mL 10% sucrose and transfer to a 50 mL falcon tube.
8. Harvest the cells by centrifugation at 5000 × *g* for 15 min (or 6000 × *g* for 10 min). Suspend cells in 1 mL 10% sucrose and transfer to a 1.5 mL micro centrifuge tube.
9. Quick-spin the suspended cells in a microfuge to pellet the cells, and remove ~0.9 mL of supernatant, leaving 0.1 mL liquid and cells in the bottom of the tube. Carefully resuspend cells with gentle pipetting and aliquot 50 μL into sterile 1.5 mL micro centrifuge tubes (*see* **Note 6**).

3.2.3 Electro-Pulse Application and Recovery

1. Add ~0.3–1.0 μg plasmid DNA, up to 5 μL in volume, to each 50 μL aliquot of competent cells. Incubate the mixtures at room temperature for 15 min.
2. Add cell/DNA mixture to the prechilled standard 1 mm electroporation cuvette (*see* **Note 7**). Tap the cuvette gently on the counter to move cells to the bottom.
3. Set the electroporator parameters: DNA transformation of *C. bescii* was successfully performed via single electric pulse (voltage: 1.8 kV and resistance: 350 Ω) using a Bio-Rad Gene Pulser with 1 mm Electroporation cuvette (*see* **Note 8**).
4. Immediately after pulsing the cells, as quickly as possible, take up and inject the cells into the bottle of preheated 10 mL LOC medium (*see* **Note 9**).
5. Incubate the bottle at 75 °C, preferably with shaking at 150 rpm from 4 to 20 h.

3.3 Select for Transformants

Currently, we are using two-step selection. First, liquid media selection is performed for simplicity and to avoid background growth on solid media. Background growth is a formidable problem, especially when using nutrient selection in *C. bescii*. Second, colony purification is performed using solid medium.

3.3.1 Liquid Medium Selection

1. One hundred microliter aliquots from the recovery cultures were periodically collected and subcultured into preheated 20 mL selective media (LOD media) bottles to select for uracil prototrophy (*see* **Note 10**).
2. Collect subcultures from recovery media at 45 min intervals for up to 3 or 4 h. One final subculture the following day is recommended.
3. Incubate these cultures in selective media at 75 °C. For shuttle/expression replicating vector transformation, expect to see growth after 30–48 h and for nonreplicating vector transformations, expect to see growth after 2–4 days (*see* **Notes 11** and **12**).
4. Screen the growing subculture bottles by PCR amplification using isolated total DNA.

3.3.2 Colony Purification on Solid Selective Media

1. Solid medium prepared by mixing an equal volume of liquid medium at a 2× concentration with autoclaved 2.0% (wt/vol) agar (Fisher Scientific). Both components should be kept in a 75 °C incubator to prevent solidification.
2. Prepare 1% agar and make aliquots of 2 mL into tubes held at 50 °C in a heating block to be used as overlay (*see* **Note 13**).
3. Mix two components for solid media and pour ~25–30 mL into each plate. Allow medium to solidify 10–20 min at room temperature.
4. Mix a 100 μL aliquot from confirmed transformant growth culture with a 2 mL 50 °C aliquot of 1% agar in heating block and pour onto the surface of plate. Swirl plate quickly to coat the surface.
5. Allow ~10 min for agar overlay to solidify. Then stack plates upside-down in an anaerobic tank. Degas with 3–4 cycles of vacuum-argon and immediately place in 75 °C incubator.
6. Isolated colonies should appear in 3–5 days (*see* **Note 14**).

3.4 Screening Purified Colonies from Solid Media

1. Take out anaerobic tank from incubator and cool at least 3 h prior to opening.
2. Start picking the colonies by using sterile toothpicks, as quickly as possible after opening the anaerobic tank.
3. Drop one sterile toothpick into each bottle of preheated selective media [21] at 75 °C.
4. Quickly seal bottles and degas headspace immediately with three cycles of vacuum-argon. Immediately place bottles in a 75 °C incubator.
5. Incubate cultures at 75 °C for 1–3 days to see growth.

6. Screen isolated cultures via "culture PCR" or gDNA extraction and PCR (*see* **Notes 15** and **16**).

3.4.1 Culture PCR

1. Remove 100 μL of each grown culture into a 1.5 mL tube.
2. Vortex vigorously.
3. Leave standing at room temperature for 15 min.
4. Use 1.0 μL (or 1:10 dilution) in a 10 μL PCR reaction.

3.4.2 Total DNA Extraction

1. Harvest cells from 10 mL of culture.
2. Use the Quick-gDNA™ MiniPrep Kit (D3007, Zymo Research) (*see* **Note 17**).
3. Run 10 μL PCR screening reactions using 1 μL template gDNA.

3.5 Protocol for Two-Step Markerless Gene Deletion Using pyrF Marker

The simplest chromosomal modification strategy is the direct replacement of a targeted region in the chromosome with a selection marker. However, the availability of selection markers is very limited in thermophiles, thus markerless chromosomal modification is considered the preferred method. In *C. bescii*, this strategy employing the *pyrF* marker was successfully utilized in several gene insertion and deletion experiments [16, 32, 34, 35, 41–44]. A schematic diagram for *ldh* targeted gene deletion is described in Fig. 2 [45].

1. Follow transformation protocol as described above to obtain recombinant strains with integration of plasmid DNA based on uracil prototrophy.
2. Confirmed transformants are inoculated into LOD medium containing 40 μM uracil, and incubated overnight at 75 °C to allow loop-out of the plasmid DNA.
3. Plate dilutions of this culture (100–1000-fold dilution) onto LOD plates containing 8 mM 5-FOA and 40 μM uracil for counter-selection.
4. Incubate in an anaerobic tank at 75 °C for 3–4 days.
5. Pick colonies as described above. Screen colonies by PCR amplification either using "Culture PCR" or total DNA isolation to confirm the targeted gene deletion.
6. Repeat the steps from #2 to #5 to obtain pure targeted gene deletion strain without contamination of wild-type allele.
7. The deletion will be verified by PCR amplification and sequence analysis of the targeted chromosome region.

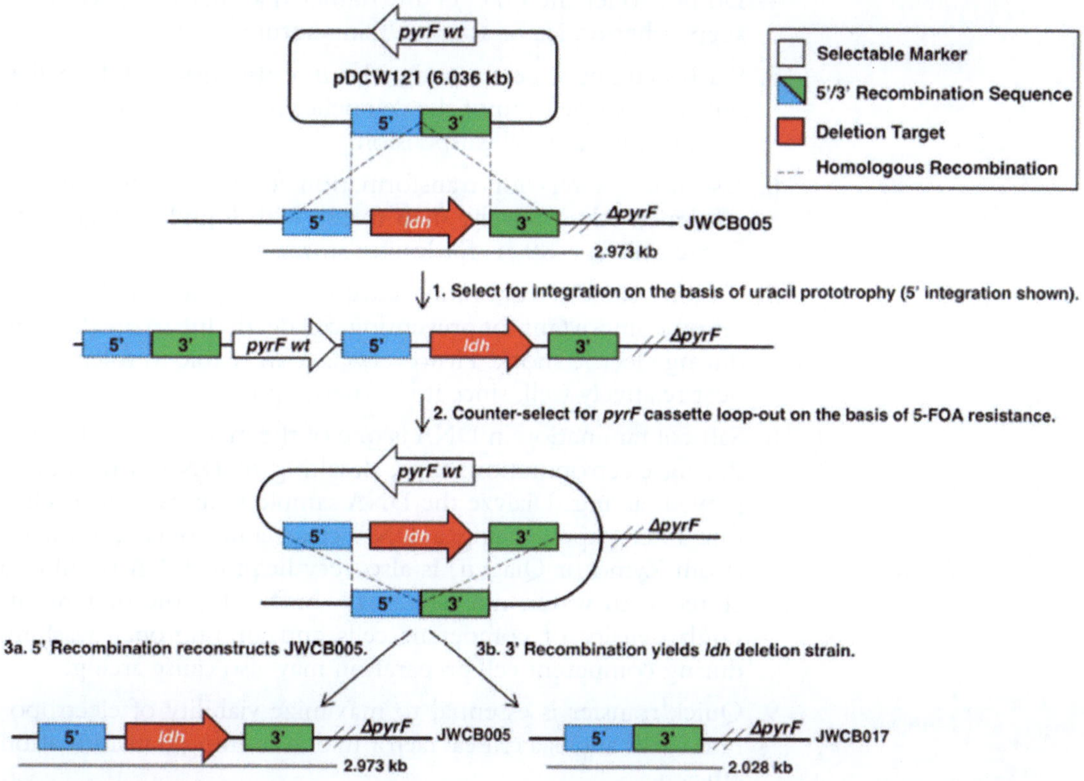

Fig. 2 Targeted gene (*ldh*) deletion using two-step markerless gene deletion with *pyrF* marker. Nonreplicating plasmid pDCW121 is constructed, containing a WT copy of the *pyrF* gene and a deletion cassette for *ldh*. The cassette contained *ldh* 5′ and 3′ flanking DNA fragments (~1.0 kb). pDCW121 is transformed into JWCB005 [*ΔpyrFA*]. Integration is selected by uracil prototrophy conferred by the *pyrF* gene. Counter-selection with 5-FOA selects for strains that underwent a second recombination event. Homologous recombination either reconstructs JWCB005 or yields the *ldh* deletion strain, JWCB017 [*ΔpyrFA/ldh*]

4 Notes

1. To allow complete methylation, an additional 2 μg of M.CbeI and 80 mM SAM may be added to the reaction for four additional hours of incubation at 78 °C.
2. This step is important to remove the salt in the DNA solution. The presence of salt will cause an arc during electroporation. This will kill the cells and significantly reduce the efficiency of transformation.
3. One 500 mL culture will generate competent cells sufficient for two transformations. Addition of amino acids to the LOD medium for competent cell preparation improves transformation efficiency.

4. Do not to let the OD get any higher than 0.15. The cells can keep either on ice or at room temperature.
5. Wash step can be performed without resuspension of the pellet, but full resuspension of the cell pellet is helpful to remove the residual media in cell suspension.
6. Use immediately for transformation for best transformation efficiency. Alternatively, snap freeze with liquid nitrogen and freeze cells at −80 °C for longer storage.
7. Chilling the electroporation cuvette on ice prior to electroporation is important for preventing cell death due to overheating during electric shock. However, *C. bescii* is able to tolerate the heat relatively well, since it is a thermophile.
8. Salt contamination in DNA is one of the main causes of arcing during electroporation; thus, desalting of DNA is helpful to prevent arcing. Dialyze the DNA sample(s) using a nitrocellulose filter paper. Using a DNA clean and concentration kit (from Zymo or Qiagen) is also very helpful. DNA should be eluted with water, not EB or TE buffer, for the best result. High density of competent cells and/or improper washing during competent cell preparation may also cause arcing.
9. Quick transfer is essential to maximize viability of electroporated cells and is a critical factor in determining transformation efficiency.
10. The subculture amount should not exceed 0.7% of LOD media to avoid background growth.
11. If transformants are not obtained, try increasing the voltage by 100 V during electroporation. High electric field often causes low transformation efficiency as it creates larger pores and too much heat. However, thermophiles are more heat resistant than mesophiles, such as *E. coli*; thus, increasing the electric strength may be helpful for transformation of large plasmid DNA.
12. If positive transformants are still not obtained, try adding an additional 4 g/L of glycine into LOD-AA medium, which is used for competent cell preparation. It causes slow growth, but this modified medium improves transformation efficiency. LOC medium with glycine also can be used. In this case, use additional care while monitoring growth due to an increased growth rate. Be careful not to overgrow cultures for competent cell preparation.
13. Using an overlay volume >2 mL will increase the number of false positive colonies for an unknown reason.
14. If colonies are not obtained after plating onto selective medium, use a freshly grown culture. The plating efficiency drops significantly when using old cultures, thus cultures more

than 2 days old should not be plated, or should be plated without dilution.

15. These techniques are very useful for initial screening to isolate transformants and can save resources and time. However, these PCR approaches do not always work very well, especially in extreme thermophiles such as *C. bescii*. This might be due to the presence of functional thermostable proteins, which are not denatured at 90 °C, being stably maintained during PCR reactions. Using thermostable DNA polymerase rather than Taq polymerase and longer pre-incubation at 98 °C might help.
16. Using freshly grown culture and more serial dilutions will also help for PCR amplification.
17. It is described as a gDNA kit, yet plasmid DNA will also be obtained using this kit.

Acknowledgments

We thank Gina L. Lipscomb for sharing *C. bescii* transformation protocols. Funding was provided by the BioEnergy Science Center (BESC) and the Center for Bioenergy Innovation (CBI), from the U.S. Department of Energy Bioenergy Research Centers supported by the Office of Biological and Environmental Research in the DOE Office of Science. This work was authored in part by Alliance for Sustainable Energy, LLC, the Manager and Operator of the National Renewable Energy Laboratory for the U.S. Department of Energy (DOE) under Contract No. DE-AC36-08GO28308. The views expressed in the article do not necessarily represent the views of the DOE or the U.S. Government. The U.S. Government retains and the publisher, by accepting the article for publication, acknowledges that the U.S. Government retains a nonexclusive, paid-up, irrevocable, worldwide license to publish or reproduce the published form of this work, or allow others to do so, for U.S. Government purposes.

References

1. Schenk LK, Kelley JH (2010) Mothering an extremely low birth-weight infant: a phenomenological study. Adv Neonat Care 10:88–97
2. Taylor MP, van Zyl L, Tuffin IM, Leak DJ, Cowan DA (2011) Genetic tool development underpins recent advances in thermophilic whole-cell biocatalysts. Microb Biotechnol 4:438–448
3. Tamakoshi M, Oshima T (2011) Genetics of thermophiles. In: Horikoshi K (ed) Extremophiles handbook. Springer, Tokyo, pp 547–566
4. Cava F, Hidalgo A, Berenguer J (2009) Thermus thermophilus as biological model. Extremophiles 13:213–231
5. Lipscomb GL, Stirrett K, Schut GJ, Yang F, Jenney FE Jr, Scott RA, Adams MW, Westpheling J (2011) Natural competence in the hyperthermophilic archaeon Pyrococcus furiosus facilitates genetic manipulation: construction of markerless deletions of genes encoding the two cytoplasmic hydrogenases. Appl Environ Microbiol 77:2232–2238
6. Koyama Y, Hoshino T, Tomizuka N, Furukawa K (1986) Genetic transformation of the

extreme thermophile Thermus thermophilus and of other Thermus spp. J Bacteriol 166:338–340

7. Sanders ME, Nicholson MA (1987) A method for genetic transformation of nonprotoplasted Streptococcus lactis. Appl Environ Microbiol 53:1730–1736
8. Wu LJ, Welker NE (1989) Protoplast transformation of *Bacillus stearothermophilus* NUB36 by plasmid DNA. J Gen Microbiol 135:1315–1324
9. Sato T, Fukui T, Atomi H, Imanaka T (2003) Targeted gene disruption by homologous recombination in the hyperthermophilic archaeon Thermococcus kodakaraensis KOD1. J Bacteriol 185:210–220
10. Fukui T, Atomi H, Kanai T, Matsumi R, Fujiwara S, Imanaka T (2005) Complete genome sequence of the hyperthermophilic archaeon Thermococcus kodakaraensis KOD1 and comparison with Pyrococcus genomes. Genome Res 15:352–363
11. Yao S, Mikkelsen MJ (2010) Identification and overexpression of a bifunctional aldehyde/alcohol dehydrogenase responsible for ethanol production in Thermoanaerobacter mathranii. J Mol Microbiol Biotechnol 19:123–133
12. Chung D, Cha M, Farkas J, Westpheling J (2013) Construction of a stable replicating shuttle vector for Caldicellulosiruptor species: use for extending genetic methodologies to other members of this genus. PLoS One 8: e62881
13. Olson DG, Lynd LR (2012) Chapter seventeen - transformation of clostridium Thermocellum by electroporation. In: Harry JG (ed) Methods in enzymology. Academic Press, Cambridge, pp 317–330
14. Maezato Y, Johnson T, McCarthy S, Dana K, Blum P (2012) Metal resistance and lithoautotrophy in the extreme thermoacidophile Metallosphaera sedula. J Bacteriol 194:6856–6863
15. Guss AM, Olson DG, Caiazza NC, Lynd LR (2012) Dcm methylation is detrimental to plasmid transformation in clostridium thermocellum. Biotechnol Biofuels 5:30
16. Chung D, Farkas J, Westpheling J (2013) Overcoming restriction as a barrier to DNA transformation in Caldicellulosiruptor species results in efficient marker replacement. Biotechnol Biofuels 6:82
17. Farkas J, Chung D, Debarry M, Adams MW, Westpheling J (2011) Defining components of the chromosomal origin of replication of the Hyperthermophilic Archaeon Pyrococcus furiosus needed for construction of a stable replicating shuttle vector. Appl Environ Microbiol 77:6343–6349
18. Tripathi SA, Olson DG, Argyros DA, Miller BB, Barrett TF, Murphy DM, McCool JD, Warner AK, Rajgarhia VB, Lynd LR, Hogsett DA, Caiazza NC (2010) Development of pyrF-based genetic system for targeted gene deletion in clostridium thermocellum and creation of a pta mutant. Appl Environ Microbiol 76:6591–6599
19. Tsoi TV, Chuvil'skaia NA, Atakishieva I, Dzhavakhishvili T, Akimenko VK (1987) Clostridium thermocellum--a new object of genetic studies. Mol Gen Mikrobiol Virusol 11:18–23
20. Mai V, Lorenz WW, Wiegel J (1997) Transformation of Thermoanaerobacterium sp. strain JW/SL-YS485 with plasmid pIKM1 conferring kanamycin resistance. FEMS Microbiol Lett 148:163–167
21. Zeldes BM, Keller MW, Loder AJ, Straub CT, Adams MW, Kelly RM (2015) Extremely thermophilic microorganisms as metabolic engineering platforms for production of fuels and industrial chemicals. Front Microbiol 6:1209
22. Chung DH, Huddleston JR, Farkas J, Westpheling J (2011) Identification and characterization of CbeI, a novel thermostable restriction enzyme from Caldicellulosiruptor bescii DSM 6725 and a member of a new subfamily of HaeIII-like enzymes. J Ind Microbiol Biotechnol 38:1867–1877
23. Yang SJ, Kataeva I, Hamilton-Brehm SD, Engle NL, Tschaplinski TJ, Doeppke C, Davis M, Westpheling J, Adams MW (2009) Efficient degradation of lignocellulosic plant biomass, without pretreatment, by the thermophilic anaerobe "Anaerocellum thermophilum" DSM 6725. Appl Environ Microbiol 75:4762–4769
24. Yang SJ, Kataeva I, Wiegel J, Yin Y, Dam P, Xu Y, Westpheling J, Adams MW (2010) Classification of 'Anaerocellum thermophilum' strain DSM 6725 as Caldicellulosiruptor bescii sp. nov. Int J Syst Evol Microbiol 60:2011–2015
25. Dam P, Kataeva I, Yang SJ, Zhou F, Yin Y, Chou W, Poole FL 2nd, Westpheling J, Hettich R, Giannone R, Lewis DL, Kelly R, Gilbert HJ, Henrissat B, Xu Y, Adams MW (2011) Insights into plant biomass conversion from the genome of the anaerobic thermophilic bacterium Caldicellulosiruptor bescii DSM 6725. Nucleic Acids Res 39:3240–3254
26. Brunecky R, Alahuhta M, Xu Q, Donohoe BS, Crowley MF, Kataeva IA, Yang SJ, Resch MG, Adams MW, Lunin VV, Himmel ME, Bomble YJ (2013) Revealing nature's cellulase

diversity: the digestion mechanism of Caldicellulosiruptor bescii CelA. Science 342:1513–1516

27. Clausen A, Mikkelsen MJ, Schroder I, Ahring BK (2004) Cloning, sequencing, and sequence analysis of two novel plasmids from the thermophilic anaerobic bacterium Anaerocellum thermophilum. Plasmid 52:131–138
28. Svetlichnyi VA, Svetlichnaya TP, Chernykh NA, Zavarzin GA (1990) Anaerocellum Thermophilum Gen. Nov Sp. Nov. an extremely thermophilic cellulolytic eubacterium isolated from hot-springs in the valley of geysers. Microbiology 59:598–604
29. Farkas J, Chung D, Cha M, Copeland J, Grayeski P, Westpheling J (2013) Improved growth media and culture techniques for genetic analysis and assessment of biomass utilization by Caldicellulosiruptor bescii. J Ind Microbiol Biotechnol 40:41–49
30. Chung D, Young J, Bomble YJ, Vander Wall TA, Groom J, Himmel ME, Westpheling J (2015) Homologous expression of the Caldicellulosiruptor bescii CelA reveals that the extracellular protein is glycosylated. PLoS One 10:e0119508
31. Kim SK, Chung D, Himmel ME, Bomble YJ, Westpheling J (2016) Heterologous expression of family 10 xylanases from Acidothermus cellulolyticus enhances the exoproteome of Caldicellulosiruptor bescii and growth on xylan substrates. Biotechnol Biofuels 9:176
32. Kim SK, Chung D, Himmel ME, Bomble YJ, Westpheling J (2017) In vivo synergistic activity of a CAZyme cassette from Acidothermus cellulolyticus significantly improves the cellulolytic activity of the C. bescii exoproteome. Biotechnol Bioeng 114(11):2474–2480
33. Kim SK, Chung D, Himmel ME, Bomble YJ, Westpheling J (2017) Engineering the N-terminal end of CelA results in improved performance and growth of Caldicellulosiruptor bescii on crystalline cellulose. Biotechnol Bioeng 114:945–950
34. Chung D, Pattathil S, Biswal AK, Hahn MG, Mohnen D, Westpheling J (2014) Deletion of a gene cluster encoding pectin degrading enzymes in Caldicellulosiruptor bescii reveals an important role for pectin in plant biomass recalcitrance. Biotechnol Biofuels 7:147
35. Young J, Chung D, Bomble YJ, Himmel ME, Westpheling J (2014) Deletion of Caldicellulosiruptor bescii CelA reveals its crucial role in the deconstruction of lignocellulosic biomass. Biotechnol Biofuels 7:142
36. Cha M, Wang H, Chung D, Bennetzen JL, Westpheling J (2013) Isolation and bioinformatic analysis of a novel transposable element, ISCbe4, from the hyperthermophilic bacterium, Caldicellulosiruptor bescii. J Ind Microbiol Biotechnol 40:1443–1448
37. Widdel F, Pfennig N (1981) Studies on dissimilatory sulfate-reducing bacteria that decompose fatty acids. I. Isolation of new sulfate-reducing bacteria enriched with acetate from saline environments. Description of Desulfobacter postgatei gen. Nov., sp. nov. Arch Microbiol 129:395–400
38. Wolin EA, Wolin MJ, Wolfe RS (1963) Formation of methane by bacterial extracts. J Biol Chem 238:2882–2886
39. Adams MW, Holden JF, Menon AL, Schut GJ, Grunden AM, Hou C, Hutchins AM, Jenney FE Jr, Kim C, Ma K, Pan G, Roy R, Sapra R, Story SV, Verhagen MF (2001) Key role for sulfur in peptide metabolism and in regulation of three hydrogenases in the hyperthermophilic archaeon Pyrococcus furiosus. J Bacteriol 183:716–724
40. Chung D, Farkas J, Huddleston JR, Olivar E, Westpheling J (2012) Methylation by a unique alpha-class N^4-cytosine methyltransferase is required for DNA transformation of Caldicellulosiruptor bescii DSM6725. PLoS One 7: e43844
41. Chung D, Cha M, Guss AM, Westpheling J (2014) Direct conversion of plant biomass to ethanol by engineered Caldicellulosiruptor bescii. Proc Natl Acad Sci U S A 111:8931–8936
42. Chung D, Cha M, Snyder EN, Elkins JG, Guss AM, Westpheling J (2015) Cellulosic ethanol production via consolidated bioprocessing at 75 degrees C by engineered Caldicellulosiruptor bescii. Biotechnol Biofuels 8:163
43. Chung D, Cha M, Young J, and Westpheling J (2014) Heterologous expression of extracellular Acidothermus cellulolyticus endoglucanase E1 in Caldicellulosiruptor bescii. In preparation
44. Cha M, Chung D, Westpheling J (2016) Deletion of a gene cluster for [Ni-Fe] hydrogenase maturation in the anaerobic hyperthermophilic bacterium Caldicellulosiruptor bescii identifies its role in hydrogen metabolism. Appl Microbiol Biotechnol 100:1823–1831
45. Cha M, Chung D, Elkins J, Guss A, Westpheling J (2013) Metabolic engineering of Caldicellulosiruptor bescii yields increased hydrogen production from lignocellulosic biomass. Biotechnol Biofuels 6:85

Chapter 3

Methods for Metabolic Engineering of *Thermoanaerobacterium saccharolyticum*

Shuen Hon, Liang Tian, Tianyong Zheng, Jingxuan Cui, Lee R. Lynd, and Daniel G. Olson

Abstract

In this work, we describe genetic tools and techniques for engineering *Thermoanaerobacterium saccharolyticum*. In particular, the *T. saccharolyticum* transformation protocol and the methods for selecting for transformants are described. Methods for determining strain phenotypes are also presented.

Key words *Thermoanaerobacterium saccharolyticum*, Metabolic engineering, Strain phenotypes

1 Introduction

Thermoanaerobacterium saccharolyticum is a Gram-positive, thermophilic, hemi-cellulolytic bacterium that was originally isolated from Yellowstone National Park from enrichments grown on xylan at pH below 4.5 [1]. A genetic system for this organism was developed in 1997, one of the first for a Gram-positive thermophile [2]. It was later discovered that this organism is naturally competent and can be transformed without electroporation [3]. Metabolic engineering of this strain began with the disruption of the lactate production pathway [4]. Additional engineering to eliminate acetate production resulted in a strain that produced ethanol at 90% of the theoretical maximum yield [5]. Subsequently, markerless gene deletion systems were developed that eliminated the constraints imposed by the limited choice of positive selection markers [6]; these tools were then used to build an ethanol producing strain of *T. saccharolyticum* that remained genetically pliable (i.e., free of genetic markers and thus amenable to subsequent modifications) after the native gene deletions [7, 8]. These genetic tools have

Shuen Hon and Liang Tian co-first authors.

Michael E. Himmel and Yannick J. Bomble (eds.), *Metabolic Pathway Engineering*, Methods in Molecular Biology, vol. 2096, https://doi.org/10.1007/978-1-0716-0195-2_3,

allowed researchers to investigate the metabolism of *T. saccharolyticum* [9–12].

The techniques for genetic manipulation of *T. saccharolyticum* have evolved between 1997 and 2017. Researchers new to *T. saccharolyticum* have therefore had to compile protocols from multiple sources, some of which may have been improved over time. This chapter aims to provide a comprehensive guide to current genetic tools and techniques for genetic modification of *T. saccharolyticum*, and to detail experimental procedures routinely used to evaluate the phenotypes of engineered strains.

2 Materials

2.1 Strains

T. saccharolyticum is available from the DSMZ culture collection as strain DSM 8691 (this strain is also known as JW/SL-YS 485 and LL1025) and was kindly given to the Lynd Lab by Dr. Juergen Wiegel. Sequence data for this strain is available from Genbank, accession number CP003184.1. Additional Illumina sequencing data for this strain is available from the Genbank sequence read archive (SRA), accession number SRA234880. Transcriptomic and proteomic data is also available for this strain [13].

Amplification of 16S ribosomal RNA (16S rRNA) by PCR and subsequent sequencing is a common method for determining strain purity [14]. When sequencing the 16S rRNAs of *T.* saccharolyticum, it is recommended to sequence from the 3′ end only (i.e., sequence using the 16S reverse primer, 5′—ACGGCTACCTTGTTACGACTT—3′) due to the presence of five copies of the 16S RNA in *T. saccharolyticum*, three of which exhibit variation in the first 250 bp of their sequences (i.e., at the 5′ end, Fig. 1). Sequencing from the 3′ end ensures that the reads cover the fully conserved regions of the five sequences. If confirmation of the 5′ end of the 16S rRNA sequence is desired, the PCR product must be subcloned to avoid sequencing a mixture of rRNA sequences.

2.2 Media and Strain Handling

1. Culture and genetic manipulation. Medium M122C has been widely used for *T. saccharolyticum*. This medium is adapted from DSMZ medium 122 [3] and its composition is described in Table 1. A slight variation on this medium recipe is called CTFUD [15] (sometimes, "modified CTFUD," although it too, is originally derived from DSMZ medium 122). The primary difference between M122C and CTFUD is that M122C uses a phosphate buffer and CTFUD uses a MOPS buffer. We recommend CTFUD for routine growth. Typically, an initial pH of 6.2 is used, unless kanamycin is being used for selection, [9–12] in which case the pH should be adjusted to 6.7 (*see* Table 3, "Special considerations").

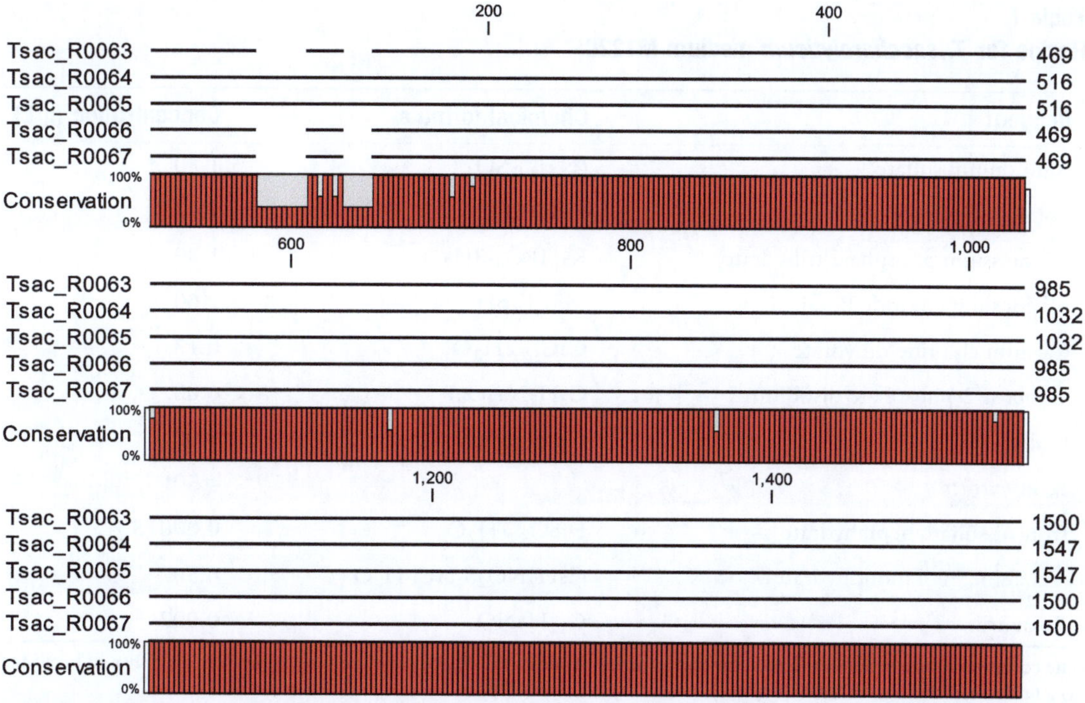

Fig. 1 An alignment of the 5 *T. saccharolyticum* 16S ribosomal RNAs. Sequences are written in the 5′ to 3′ direction. The gene number of the rRNA sequence is based on the Genbank CP003184 sequence. The bar graph under the sequence indicates the average conservation at that position

2. Carbon source. *T. saccharolyticum* can grow on a wide range of sugars as well as starch and xylan. For routine culture, cellobiose is commonly used. Glucose, xylose, arabinose, maltodextrin [8], mannose, and galactose can also be used [16, 17]. One benefit of xylose is that it is not a common substrate for other organisms, and thus is sometimes used when contamination is a concern. Xylose rapidly degrades at high temperature, though, so media prepared with xylose should not be autoclaved for more than 30 min and should be promptly removed from the autoclave after sterilization.

Certain selection agents (*see* Subheading 2.3) require a chemically defined medium to minimize the appearance of background colonies. In this case, the yeast extract in M122C (Table 1) should be replaced with RPMI 1640 vitamins (Sigma R7131, Sigma Aldrich) and 1× minimal essential medium (MEM) amino acids (Sigma M5550, Sigma Aldrich) [7]; 20 mL of each solution is used per 1 L of medium. The resulting medium is known as "M122 defined."

Another defined medium used for analyses of fermentation end products, MTC-6, is adapted from defined *C. thermocellum*

Table 1
Recipe for *T. saccharolyticum* medium M122C

Reagent	Chemical formula	Concentration (g/L)
Ammonium sulfate	$(NH_4)_2SO_4$	1.30
Potassium phosphate	KH_2PO_4	1.43
Dipotassium phosphate trihydrate	$K_2HPO_4 \cdot 3H_2O$	1.80
Magnesium chloride hexahydrate	$MgCl_2 \cdot 6H_2O$	2.60
Calcium chloride dihydrate	$CaCl_2 \cdot 2H_2O$	0.13
Glycerol-2-phosphate disodium	$C_3H_7Na_2O_6P$	6.00
Cellobiose	$C_{11}H_{22}O_{11}$	5.00
Yeast extract		4.50
Iron(II) sulfate heptahydrate	$FeSO_4 \cdot 7H_2O$	0.00013
L-cysteine HCl monohydrate	$C_3H_7NO_2S \cdot HCl \cdot H_2O$	0.50
Resazurin	$C_{12}H_7NO_4$	0.002

The compounds are listed in the recommended order of addition. Where solid medium is desired, agar should be added to a final concentration of 1.0% (wt/vol)

medium, MTC-5 [18]; the composition of MTC-6 is described in Table 2.

Freezer stocks of *T. saccharolyticum* cultures (grown either in M122C, M122 defined, or MTC-6) intended for immediate or short-term use can be stored at −80 °C without the need for cryoprotectant. For long-term storage (e.g., in a strain collection), it is recommended to add dimethyl sulfoxide (DMSO) as a cryoprotectant to a final concentration of 5% v/v [3]. The longevity of the long-term storage cultures can be further improved by storing the strains in sealed serum vials (serum vial: Wheaton catalog number 223738; stopper: Chemglass Life Sciences catalog number CLS4209-14, aluminum seal: Chemglass Life Sciences catalog number CLS-4209-12) that have been purged with 100% nitrogen gas.

It is also worth noting that cultures of *T. saccharolyticum* strain DSM 8691 are prone to autolysis; suggested causes include insufficient magnesium in the culture medium [19] and harvesting cells from cultures that have been grown beyond stationary phase [20]. For the latter reason, when it is desired to maintain cell viability (i.e., for subcultures and freezer stocks) the culture should be monitored to ensure that cells are harvested in mid-exponential phase (OD_{600} usually between 0.4 and 1.0 when grown on 5 g/L sugar). We have also observed that lysis behavior is associated with high growth rates, followed by depletion of single sugar carbon sources. Therefore, another way to avoid lysis is to grow *T. saccharolyticum* on mixed sugars, or at a slower growth rate.

Table 2
Recipe for defined medium, MTC-6

Stock Solution	Chemical Name	Chemical Formula	Concentration (g/L)
A	Cellobiose	$C_{11}H_{22}O_{11}$	Variable[a]
	MOPS sodium salt	$C_7H_{14}NNaO_4S$	9.25
B	Potassium citrate monohydrate	$C_6H_5O_7K_3$	2
	Citric acid monohydrate	$C_6H_8O_7{\cdot}H_2O$	1.25
	Sodium sulfate	Na_2SO_4	1
	Potassium phosphate	KH_2PO_4	1
	Sodium bicarbonate	$NAHCO_3$	2.5
C	Ammonium chloride	NH_4Cl	2
D	Magnesium chloride hexahydrate	$MgCl_2{\cdot}6H_2O$	1
	Calcium chloride dehydrate	$CaCl_2{\cdot}2H_2O$	0.2
	Iron(II) chloride hexahydrate	$FeCl_2{\cdot}6H_2O$	0.1
	L-cysteine HCl monohydrate	$C_3H_7NO_2S{\cdot}HCl{\cdot}H_2O$	1
E	Pyridoxamine dihydrochloride	$C_8H_{12}N_2O_2{\cdot}2HCl$	0.02
	P-aminobenzoic acid	$C_7H_7NO_2$	0.004
	D-biotin	$C_{10}H_{16}N_2O_3S$	0.004
	Vitamin B12	$C_{63}H_{88}CoN_{14}O_{14}P$	0.002
	Thiamine hydrochloride	$C_{12}H_{17}ClN_4OS{\cdot}HCl$	0.004
F (trace minerals)	Manganese(II) chloride tetrahydrate	$MnCl_2{\cdot}4H_2O$	0.0005
	Cobalt(II) chloride hexahydrate	$CoCl_2{\cdot}6H_2O$	0.0005
	Zinc chloride	$ZnCl_2$	0.0002
	Copper(II) chloride dehydrate	$CuCl_2{\cdot}2H_2O$	0.00005
	Boric acid	H_3BO_3	0.00005
	Sodium molybdate dehydrate	$Na_2MoO_4{\cdot}2H_2O$	0.00005
	Nickel(II) chloride hexahydrate	$NiCl_2{\cdot}6H_2O$	0.00005

[a]5 g/L is the most commonly used concentration of cellobiose for standardized fermentations. If more than 20 g/L of cellobiose is used, the concentrations of the components in solution E should be doubled

2.3 Selection Markers

Several positive and negative selective agents are available for use in *T. saccharolyticum* as summarized in Table 3.

Kanamycin and erythromycin are two antibiotics that can be used for positive selection in *T. saccharolyticum* [5]. Kanamycin is generally preferred because it gives lower background and it functions in conditions more similar to the optimal growth conditions for *T. saccharolyticum* (i.e., temperature and pH) [5] (Table 3).

Haloacetates (i.e., chloroacetate and fluoroacetate) and 5-fluoroorotic acid (5-FOA) were the first negative selective agents developed for *T. saccharolyticum* [6] after the corresponding endogenous selection markers had been deleted (Table 3). Chloroacetate is the preferred haloacetate (instead of fluoroacetate) due to its lower toxicity to humans. However, this selection requires chemically defined medium that reduces the rate of cell growth. More recently, a thymidine kinase-based negative selection marker (where 5-fluoro-2′-deoxyuridine (FUDR) is the negative selection

Table 3
List of known selection agents that can be used to genetically engineer *T. saccharolyticum* with

Selection agent	CAS number	Selection type	Stock concentration (mg/mL)	Working concentration (μg/mL)	Solvent	Selection marker gene (Accession number)	Special considerations	References
Kanamycin sulfate	25389-94-0	Positive	50	200	milliQ water	*htk* (KX272604)	The growth medium pH should be adjusted to 6.7	[5]
Erythromycin	114-07-8	Positive	10	10	2 M HCl or ethanol	*erm* (MF503103)	Selection temperature of 50 °C	[5]
Sodium chloroacetate	3926-62-3	Negative	11.65	23.3	milliQ water	*Tsac_1744-1745* (CP_003184)	Requires *pta-ack* (*Tsac_1744-1745*) deletionSelection should be performed on M122 defined medium, pH 6.0	[6]
5-fluoroorotic acid (5-FOA)	703-95-7	Negative	100	100	Dimethyl sulfoxide	*Tsac_1704* (CP_003184)	Requires deletion of *pyrF* (*Tsac_1704*) pH of medium should be 5.0	[6]
5-fluoro-2′-deoxyuridine (FUDR)	50-91-9	Negative	20	10–20	milliQ water	*tdk* (KX272604)	deletion prior to its use as a negative selection marker	[12]

agent) has been developed for use in *T. saccharolyticum* in markerless genetic modification (Subheadings 3.2 and 3.3) [12].

Due to their reliability and ease of use, kanamycin and FUDR are the recommended positive and negative selection markers, respectively (Table 3, "Special considerations").

2.4 T. saccharolyticum Transformation Protocol

All steps are to be performed under strict anaerobic conditions (5% H_2, 10% CO_2, 85% N_2).

1. Inoculate 10 mL of M122C medium (Table 1) with 20 μL of *T. saccharolyticum* freezer stock. Culture to mid-exponential phase (OD_{600} between 0.4 and 0.8).
2. Dilute the culture to an OD_{600} of less than 0.1. *Note*: natural competence seems to occur during the early stages of growth, so the goal of this dilution is to allow for the maximum number of generations before mid-log phase in the presence of DNA.
3. Mix 1 mL of the diluted culture with 0.25–3 μg of DNA in a sterile tube.
4. Grow the mixture of DNA and cells until it reaches mid-exponential phase (OD_{600} of 0.4–0.8 for wild-type *T. saccharolyticum*). Culturing past mid-exponential phase often results in significantly decreased transformation efficiency, possibly due to cell lysis.
5. Perform three 1:100 serial dilutions of the mid-exponential phase culture to create 10^{-2}, 10^{-4}, and 10^{-6} dilutions. Separately plate 200 μL of the competent cell culture and the three dilutions.
6. Prepare 30 mL of melted M122C agar with a selective agent (Table 3), and then add 7 mL of molten M122C agar to each quadrant of a four-well compartmentalized petri dish (e.g., Fisher Scientific catalog number FB087582). Note that some selective agents require modifications to the medium (Table 3).
7. To mix the agar and cells, pipette the contents of each quadrant several (~10) times, taking care not to leave behind bubbles that will hinder colony picking once the agar solidifies. A 1 mL pipette is particularly well suited to this task. Allow the agar to solidify at room temperature for 30 min.
8. Incubate the plates at 55 °C, unless otherwise specified in Table 3. Colonies should be visible within 1–5 days.
9. Colonies are picked by aspirating a small volume (1–10 μL) of the colony into a pipette tip. We typically prefer to pick colonies embedded in the agar (as opposed to surface colonies) to avoid cross-contamination (i.e., during incubation, water may condense on the surface of the plate. Routine handling of the plates may cause those water droplets to slide across the surface of the plate, cross-contaminating surface colonies). The picked

colony should be resuspended (briefly) in sterile molecular grade water or culture medium. An aliquot of the colony suspension is then taken for analysis by PCR, and the rest is used to start a liquid culture of the colony.

10. Note: when performing the negative selection, it is recommended to also set up a plate that lacks the negative selective agent as a control, so that the selection efficiency can be determined.

3 Genetic Tools

3.1 Theoretical Considerations

3.1.1 Design Chromosome Modification Constructs

There are several different kinds of chromosomal modifications that can be performed in *T. saccharolyticum*, and a variety of ways to achieve these modifications. Here we will describe the key decisions that are involved in the construct design process. These decisions can be combined to yield many chromosome modification strategies. We provide a few detailed examples below.

3.1.2 Insertion Versus Deletion

The first decision is what to modify. In general, chromosomal modifications can be either deletions, insertions, or a combination of a deletion and an insertion. Making chromosomal modifications in *T. saccharolyticum* depends on homologous recombination between the chromosome and the modification construct. Therefore, the choice of homology region is determined by modification type. For a deletion, the upstream flank (5′ flank) and downstream (3′ flank) are chosen such that the region between the flanks is the region that will be deleted. Note that the entire sequence of 5′ and 3′ flanks will remain after the deletion. For an insertion, the 5′ and 3′ flanks are chosen to be adjacent to each other. The boundary between the 5′ and 3′ flanks identifies the site of the insertion. When making both an insertion and a deletion, the 5′ and 3′ flanks are designed as for a deletion. Homology regions typically contain 500–1000 bp of homology to the *T. saccharolyticum* genome.

3.1.3 Single Versus Dual Constructs

The second decision is whether to use the single-construct or dual-construct method. In the single-construct method, three homology regions are used (5′ flank, 3′ flank and internal or “int” region). The 5′ flank and 3′ flank are chosen as described in the previous paragraph. The int region is internal to the gene that is being deleted (in the case of a deletion) (*see* Fig. 5). For a gene insertion (without deletion), a second copy of the 3′ flank is used in place of the int region. A selection cassette, consisting of a positive and negative selectable marker, is inserted between the 3′ flank and int region of the construct. The advantage of the single-construct method is that only one construct and one transformation event is needed per genetic modification. A final advantage is that this method is more similar to the one used for *Clostridium thermocellum* [15], so for people working with both organisms, this method is preferred. The disadvantages are that the construct is more difficult to build because there are more parts to assemble (3–6), and sometimes these parts involve repetitious DNA sequence (i.e., if the construct has two copies of the 3′ flank). Another disadvantage is that the additional homology flanking regions allow for a greater number and diversity of unintended recombination events, which can lead to confusion when trying to confirm the intermediate recombination steps by PCR.

In the dual-construct method, only a 5′ flank and 3′ flank are needed, but two separate DNA constructs are prepared. The first construct is used to replace the region between the 5′ and 3′ flanks with a selection cassette. This construct has a 5′ flank—selection cassette—3′ flank design. The second construct is used to eliminate the selection cassette to either produce an unmarked deletion or a targeted insertion. For unmarked deletions, the construct is made with adjacent 5′ and 3′ flanks. For an insertion, the construct is made with the design 5′ flank—insertion—3′ flank. The advantage of the dual-construct method is that the individual constructs have fewer pieces (2–4), and the presence of only two homology regions means there are fewer options for unintended recombination events. Although two transformation events are required (one for each construct), since *T. saccharolyticum* is relatively easy to transform, this does not add much time or complexity to the experiment.

Both the single-construct [12, 21, 22] and the dual-construct [6] methods are currently in use in our lab.

3.1.4 Circular Versus Linear DNA

Constructs for chromosomal modification can be made of either circular DNA (i.e., plasmid) or linear DNA (i.e., PCR product). For chromosomal integration, circular DNA constructs do not include a *T. saccharolyticum* origin of replication. A benefit of using a circular plasmid DNA is that it can be maintained and amplified in *E. coli*, with a high degree of fidelity. A benefit of using linear DNA is that it can be generated quickly, and in vitro. Building DNA constructs in-vitro avoids limitations due to toxicity in *E. coli* that sometimes is experienced with circular plasmids. Another benefit of linear DNA is that it adds an additional selection for the double recombination event, since a single recombination event would result in a lethal chromosomal break without the second recombination event. Both linear and circular DNA can be used to make unmarked gene deletions.

3.2 Example 1: Gene Deletion Using Dual-Construct Method with Linear DNA

This example describes the construction of an unmarked deletion of the *pyrF* gene in *T. saccharolyticum* using the dual-construct method. This is similar to the protocol described by Shaw et al. using a combination of circular and linear DNA, [6] but here we use only linear DNA.

3.2.1 Construct Design

The first construct is used to delete the target gene (*ldh*) and replace it with the selection cassette. The selection cassette uses the *pta* and *ack* genes for negative selection with chloroacetate, and the *kanR* gene for positive selection with kanamycin (Table 3). Thus, the first construct has a 5′ flank—selection cassette—3′ flank pattern (Fig. 2).

The second construct is used to remove the selection cassette. In this construct, only the 5′ and 3′ flanks are needed (Fig. 2).

3.2.2 Detailed Protocol

1. Follow transformation protocol as described above (Subheading 2.4), using the first DNA construct.
2. Screen colonies by PCR to confirm the presence of the selection cassette, and to confirm that the allelic replacement has been successful. (Fig. 3, panel c). Three pairs of primers are used for confirmation PCRs: one pair that binds to the chromosome external to the homology flanks, another that binds to an internal region of the selection marker (Fig. 3, panel c), and a third that binds to the region of the gene of interest targeted for deletion (Fig. 3, panel a).
3. At this point, the gene has been disrupted, and the protocol can be stopped. The following steps in the protocol are used to remove the selection cassette so that it can be used for another round of selection.
4. Perform a second transformation with the second DNA construct (Subheading 2.4).
5. Screen colonies by PCR. At this stage, two pairs of primers are used: one pair that binds to the chromosome external to the homology flanks, and a second pair that binds to the region of the gene of interest targeted for deletion (Fig. 3, panels e–f).

3.3 Example 2: Gene Deletion Using Single-Construct Method with Circular DNA

3.3.1 Construct Design

The single-construct method uses the same 5′ flank and 3′ flanks that were used in the dual-construct method. Additionally, there is a third region that is homologous to a region internal to the targeted gene, called the "int" region. This internal homology region should also be between 500 and 1000 bp in size, although if the target gene is small (e.g., less than 200 bp), a second copy of the 3′ flank can be used in place of the internal region.

The selection cassette for a markerless deletion plasmid comprises both a positive and negative selection marker. In this

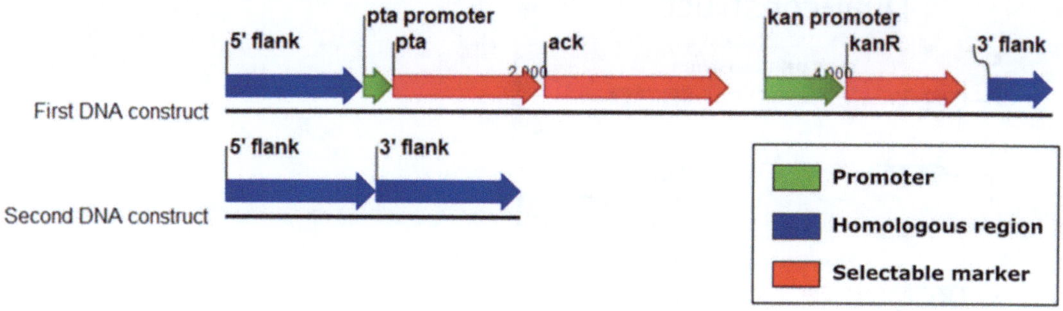

Name	Type	Region	Orientation	Description
5′ flank	Homology region	1-904 (1st) 1-992 (2nd)	Fwd	5′ (upstream) flanking region of the gene target
pta promoter	Promoter	908-1101	Fwd	Promoter for *pta* gene
pta	CDS	1102-2088	Fwd	Phosphate acetyltransferase, confers sensitivity towards haloacetates
ack	CDS	2106-3317	Fwd	Acetate kinase, confers sensitivity towards haloacetates
kan promoter	Promoter	3549-4074	Fwd	Promoter for the *kanR* gene
kanR	CDS	4087-4881	Fwd	Aminoglycoside phosphotransferase, provides resistance to kanamycin
3′ flank	Homology region	5034-5466 (1st) 993-1951(2nd)	Fwd	3′ (downstream) flanking region of the gene target

Fig. 2 Diagram of DNA constructs for dual-constructs method; the nucleotide sequence is available from Genbank (MF688931 and MF688932)

example, we use *htk* as the positive marker and *tdk* as the negative marker (Fig. 4).

Flanks should be placed in the order: 5′ flank—3′ flank—selection cassette—int region (Fig. 4). Note that the orientation of homologous flanks must be in the same direction (relative to their orientation on the chromosome) or the desired recombination events will not occur. Plasmid pLT_26 (Fig. 4) (GenBank accession number KX272604) is an example of a markerless deletion plasmid.

3.3.2 Detailed Protocol

1. Follow the transformation protocol from Subheading 2.4, using a positive selection (i.e., kanamycin) to select for the presence of the selection cassette on the genome (Fig. 5, panels a–d).

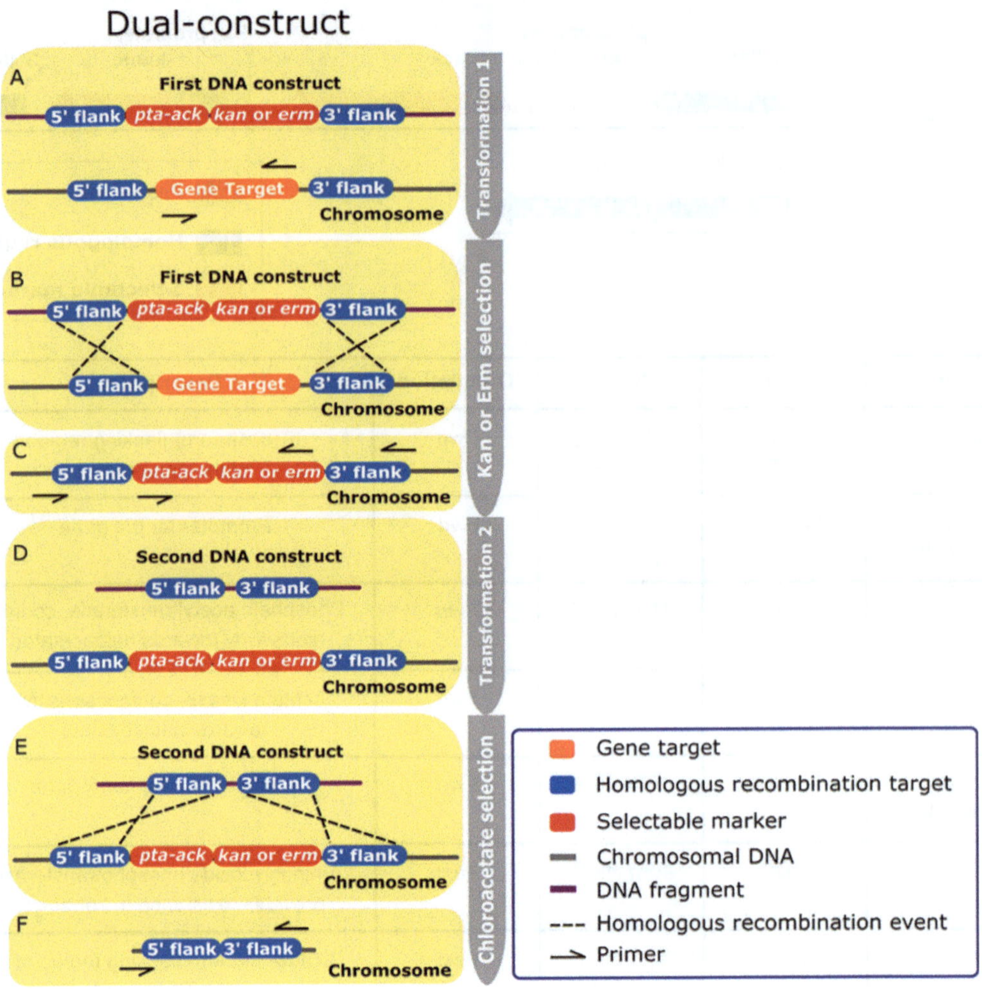

Fig. 3 Selection scheme diagram for dual-construct method

2. Screen colonies by PCR to confirm presence of the marker gene and first two recombination events. At this stage, the strain is referred to as merodiploid due to the presence of two copies of the 3′ flank (Fig. 5, panel d).
3. Inoculate positive colonies into liquid medium containing the positive selection agent (Table 3).
4. When the culture has reached mid-exponential phase, plate four 100-fold serial dilutions on solid medium containing the negative selection agent, paying attention to medium requirements if any (Table 3).
5. Colonies should appear in 1–5 days.
6. Screen colonies by PCR to confirm that the third recombination event has occurred (Fig. 5, panel e and f). At this point, the targeted gene deletion has been made and the marker has been removed.

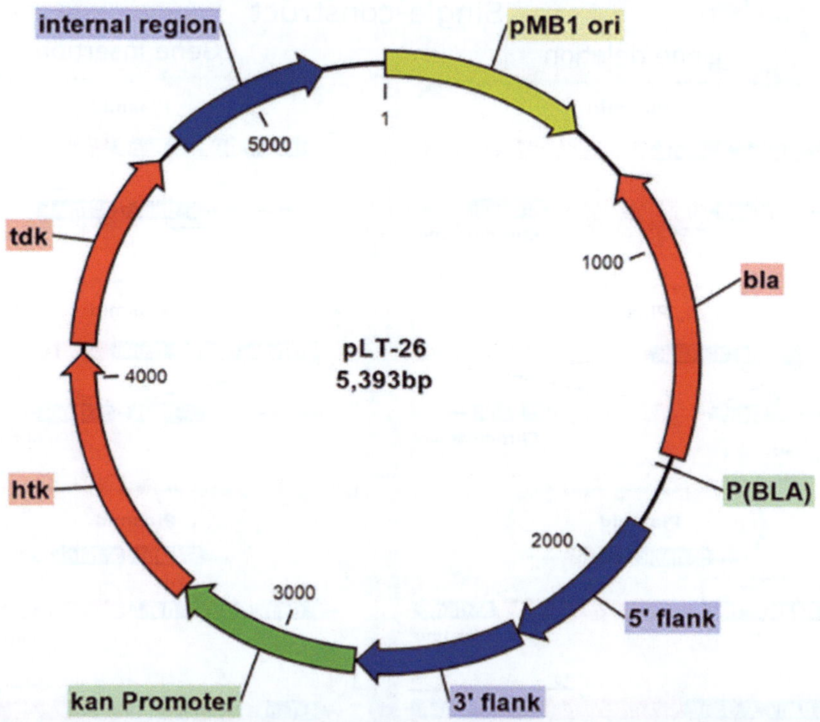

Name	Type	Region	Orientation	Description
pMB1 ori	Replication origin	1-589	Fwd	pUC19 origin of replication for plasmid propogation in *E. coli*
bla	CDS	763-1620	Rev	Beta lactamase, provides resistance to ampicillin
bla promoter	Promoter	1620-1655	Rev	Promoter for *bla* gene
5' flank	Homology region	1837-2306	Fwd	5' (upstream) flanking region of the gene target
3' flank	Homology region	2307-2776	Fwd	3' (downstream) flanking region of the gene target
kan promoter	Promoter	2781-3311	Fwd	Promoter for *htk* gene
htk	CDS	3321-4058	Fwd	Thermostable kanamycin nucleotidyltransferase, provides resistance to kanamycin
tdk	CDS	4097-4678	Fwd	Thymidine kinase from *Thermoanaerobacterium saccharolyticum*
internal region	Homology region	4757-5218	Fwd	Internal region to the gene target

Fig. 4 Diagram of plasmid pLT_26 for creating an unmarked deletion. The nucleotide sequence is available from Genbank (accession number KX272604). Genetic features are described in the table

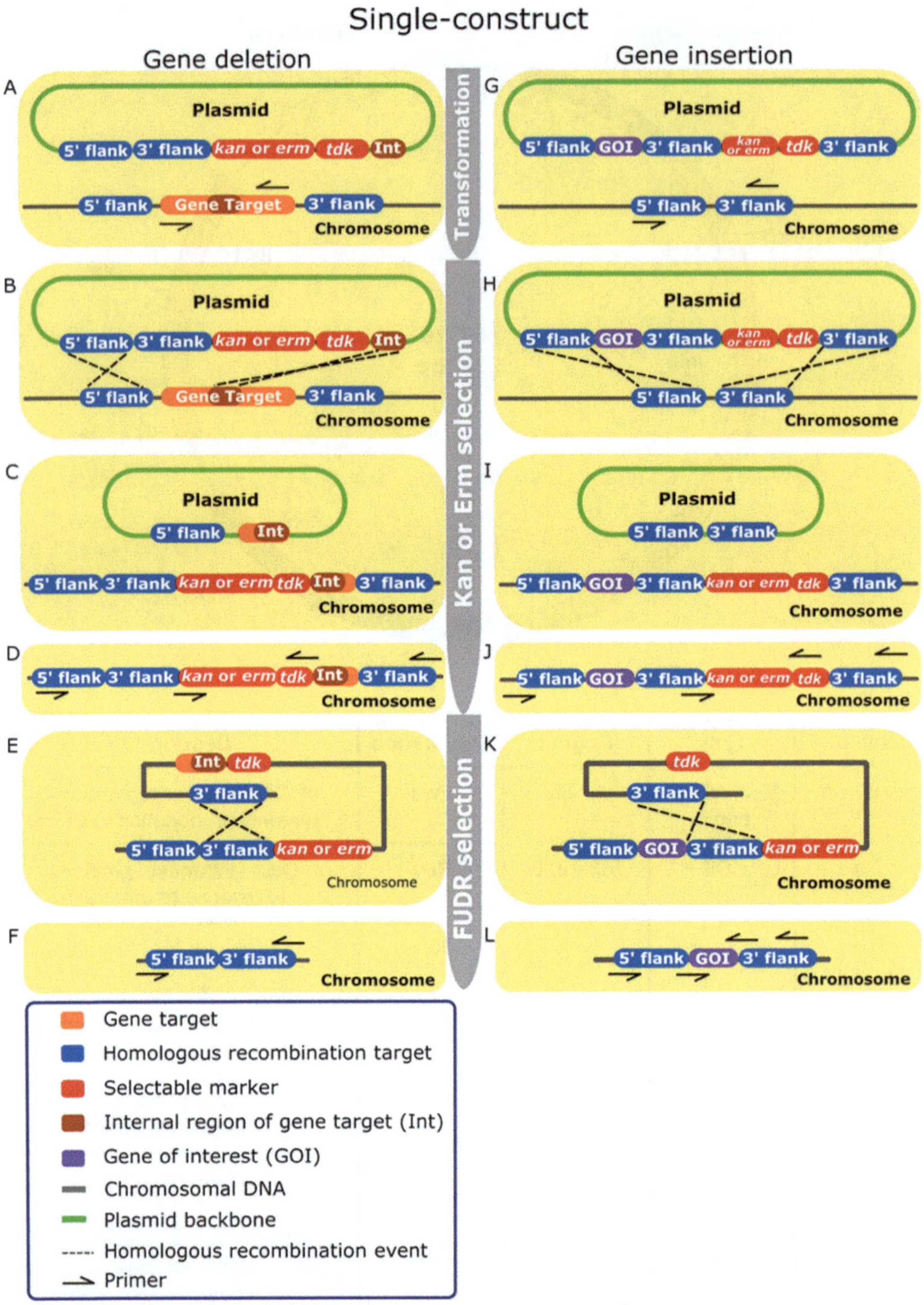

Fig. 5 Selection scheme diagram for single-construct method, gene deletion (panels **a–f**) and gene insertion (panels **g–l**)

7. Plate dilutions of the culture with the unmarked deletion genotype without selection to isolate individual colonies. This final round of colony purification is recommended to ensure purity of colonies picked from solid plates. It is also recommended to re-screen colonies from this plating by PCR, as was done in **step 6**.

3.4 Example 3: Gene Insertion

Due to the ease of chromosomal integration in *T. saccharolyticum*, gene expression is typically performed by integrating the gene onto the chromosome. Chromosomal expression is generally more stable than plasmid-based expression. The pMU131 plasmid (pLL1118 in the Lynd Lab collection, also available from AddGene, #66811) uses the cryptic replicon from *Thermoanaerobacterium saccharolyticum* strain B6A (ATCC 49915) and has been shown to replicate in *T. saccharolyticum* JW/SL-YS485, but has not been extensively tested for use in gene expression experiments [23].

3.4.1 Construct Design

The principles of gene replacement are like those of gene deletion. The only difference is that the gene to be introduced (gene of interest) is inserted between the 5′ flank and 3′ flank (Fig. 6). In the dual-construct design, the first construct is not changed. The second construct is modified to have the following sequence order: 5′ flank—gene of interest—3′ flank. In a single-construct design, the order of sequences is: 5′ flank—gene of interest—3′ flank—selection cassette—int region.

3.4.2 Detailed Protocol

The transformation and screening protocols are the same as previously described for either dual-construct (Subheading 3.2.2) or single-construct (Subheading 3.3.2) designs. When performing gene insertions, we typically perform an additional PCR to confirm the presence of the insertion. Fig. 5 (panels g–l) shows an example of using the single-construct method to perform an insertion.

3.5 Promoters

There has not yet been a systemic characterization of promoters for expressing genes in *T. saccharolyticum*. However, the kanamycin promoter (Subheadings 3.1 and 3.2) is routinely used to drive expression of selection cassettes. In addition, the *C. thermocellum cbp_2* promoter [24] has been successfully used to express the urease operon from *C. thermocellum* in *T. saccharolyticum* [7]. The native *xynA* (*Tsac_1459*) promoter can also be used, [22, 25] with the extra feature that expression of a gene under the transcriptional control of the *xynA* promoter can be selectively induced by culturing the strain with xylose as the carbon source [25].

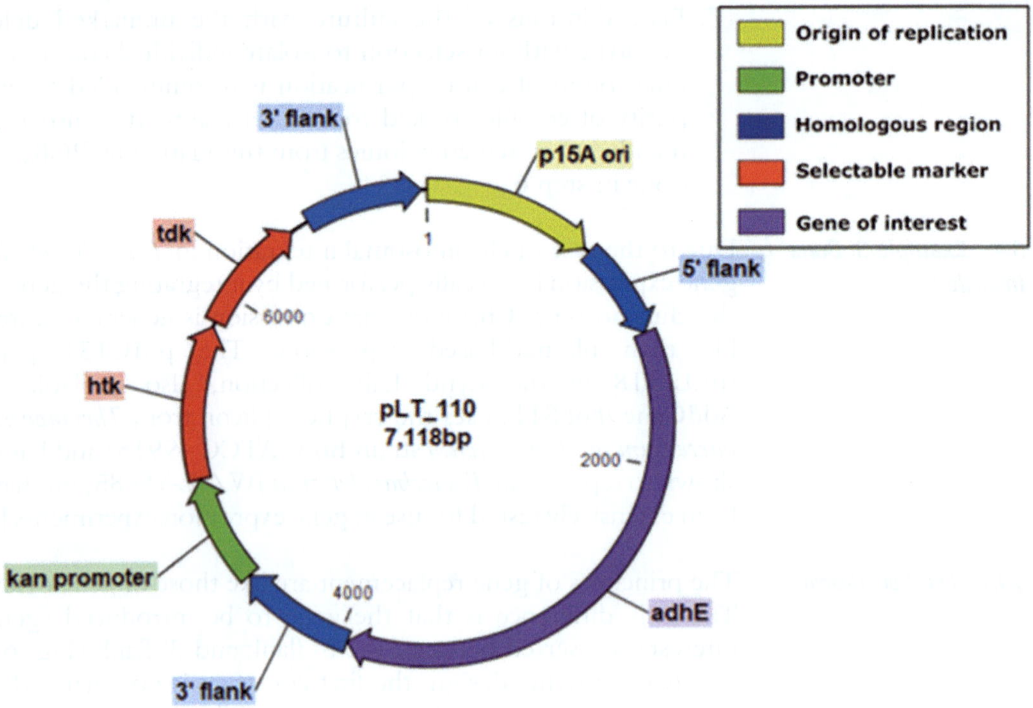

Name	Type	Region	Orientation	Description
p15A ori	Replication origin	1-828	Fwd	p15A origin of replication for plasmid propogation in *E. coli*
5' flank	Homology region	856-1335	Fwd	5' (upstream) flanking region of the insertion site
adhE	CDS	1336-3957	Fwd	Bifunctional alcohol and aldehyde dehydrogenase (Gene of interest)
3' flank	Homology region	3958-4528	Fwd	3' (upstream) flanking region of the insertion site
kan promoter	Promoter	4533-5063	Fwd	Promoter for *htk* and *tdk* genes
htk	CDS	5064-5825	Fwd	Thermostable kanamycin nucleotidyltransferase, provides resistance to kanamycin
tdk	CDS	5849-6430	Fwd	Thymidine kinase from *Thermoanaerobacterium saccharolyticum*
3' flank	Homology region	6522-7092	Fwd	Internal region to the gene target

Fig. 6 Diagram of plasmid pLT_110 for insertion of a gene of interest (in this case *adhE*). The nucleotide sequence is available from Genbank (accession number MF818012). Genetic features are described in the table

4 *T. saccharolyticum* Phenotype Analyses

4.1 Fermentation Analyses

4.1.1 Defined Medium MTC-6

For analyzing fermentation performance, it is desirable to use a chemically defined medium. Complex additives can contribute to the production of excess fermentation products that interfere with yield calculations. MTC (*m*edium for *t*hermophilic *c*lostridia) was developed by David Hogsett to allow fermentation of up to 50 g/L carbohydrate [26]. MTC-6 is a slight variation of the Hogsett formulation. The differences are that MTC-6 medium includes 4 mg/mL thiamine hydrochloride, urea is replaced with ammonium chloride as the nitrogen source, and the starting pH of the medium is 6.2 instead of 7.4. Table 2 gives the complete recipe for MTC-6. The components of MTC-6 are often prepared as separate stock solutions (denoted A–F). These stock solutions can be prepared and stored under nonsterile conditions and are stable for several months. If sterilization is desired, it should be done via filter sterilization (0.2 μm pore size filter) and not by autoclaving. Solution A is prepared fresh, while solution B is made at a 25× concentration. Solutions C–E are made at 50× concentrations, and solution F at 1000× concentration. All stock solutions are to be stored under anaerobic conditions (sealed serum bottles with the headspace purged with 100% nitrogen). Solutions B and C can be stored at room temperature, while solutions D–F should be stored at 4 °C. Solution E should also be stored away from light, which can be done by covering the bottle with aluminum foil.

4.1.2 Standardized Fermentation Conditions

Standardized fermentations are done in sealed 150 mL serum bottles, with a 50 mL fermentation volume and 100 mL headspace. The serum bottle headspace is purged with 100% nitrogen gas. Prior to inoculation, the bottles are depressurized to atmospheric pressure. A 2% v/v inoculum is added to the bottles. After inoculating, the bottles are incubated in a 55 °C shaker with shaking at 180 rpm. Fermentations are allowed to proceed for 72 h, after which they are analyzed [12].

4.1.3 Analyses of Fermentation Products

Fermentation gas products (hydrogen and carbon dioxide) are analyzed with an SRI 310C gas chromatograph (SRI Inc.), with nitrogen as the carrier gas (flow rate of 8.2 mL/min). The column oven temperature is maintained at 151 °C [12].

Fermentation products in the broth are quantified by high-pressure liquid chromatography (HPLC) with refractive index and UV detection, using an Aminex HPX-87H column (Bio-Rad, Hercules, CA), with 5 mM sulfuric acid solution as the eluent [27]. Fermentation samples are first centrifuged at >20,000 × *g* for 5 min; the supernatant is then acidified with 10% v/v sulfuric acid in a 20:1 ratio of supernatant to sulfuric acid and vortexed to mix. The purpose of acidification is to precipitate protein. If the sample was from a fermentation with a very high (>50 g/L) initial substrate concentration, the acidified sample should be centrifuged again. The acidified sample supernatant is then filtered through 0.2 μm nylon filter before sample analysis.

Pellet carbon and pellet nitrogen (used to determine cell biomass) can be measured with a Shimadzu TOC-V CPH elemental analyzer with TNM-1 and ASI-V modules (Shimadzu Corp., Columbia, MD) [27]. A 1 mL fermentation sample is first centrifuged (>20,000 × *g* for 5 min), the supernatant is removed, and the sample is washed twice with 1 mL milliQ water per wash.

Extracellular amino acids are measured via a derivatization method (Accq-Tag Chemistry kit, catalog number WAT052875, Waters Corporation, Milford, MA) coupled with HPLC with fluorescence detection, using an excitation wavelength of 250 nm and emission wavelength of 395 nm. Sample preparation is performed according to manufacturer's instructions.

4.2 Enzyme Assays

Although detailed descriptions of assay conditions for various enzyme assays have been published, [10, 11, 21, 22] there are some challenges specific to anaerobic assays that warrant discussion. Figure 7 shows the setup for maintaining both anaerobic and thermophilic conditions.

It is also recommended that all plastic consumables such as tubes and pipette tips be kept in the anaerobic chamber for at least 48 h prior to use in order to allow oxygen to diffuse out of the plastic.

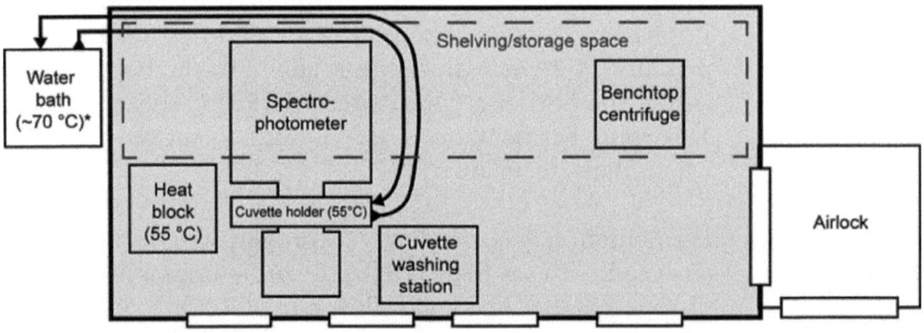

Fig. 7 Example layout in the interior of an anaerobic chamber that has been set up for anaerobic enzyme assays. The spectrophotometer is placed inside the chamber. Temperature control of the cuvette is maintained by a water bath outside the chamber, which is connected to the eight-place cuvette holder. We have also used a cuvette holder with Peltier temperature control; however, this device is limited to a single cuvette. A cuvette washer is a convenient addition. It can be purchased from FireflySci (product catalog number P65S) and is paired with a vacuum flask and portable vacuum pump. The spectrophotometer used in this setup is the Agilent 8435 model. *The temperature of the water bath should be adjusted to account for heat loss between the water bath and the cuvette holder. The user is advised to titrate the temperature set point of the water bath until the temperature of the cuvette holder stabilizes at 55 °C

4.2.1 Preparing Cell-Free Extracts

T. saccharolyticum cells can be lysed by either sonication or with lysozyme, with the latter method being more commonly reported; both methods should be performed under anaerobic conditions. To obtain cell-free extracts for enzymatic analyses, *T. saccharolyticum* cells are first thawed (if previously stored at −80 °C) and resuspended in a lysis buffer that contains 1× BugBuster reagent (EMD Millipore catalog number 70584) at pH 7.0. One μL (~30,000 U) of Ready-Lyse™ Lysozyme solution (Epicentre Biotechnologies catalog number R1804M) is added to the cell suspension and gently mixed by inverting the tube several times. The mixture is left to incubate at room temperature for 10 min; cell lysis can be observed by an increase in viscosity and decrease in opacity. One μL (~2500 U) of DNase I (Thermo Scientific catalog number 90083) is then added to the mixture to decrease the viscosity. Un-lysed cells and debris are pelleted by centrifugation at >12,000 × *g* for 5 min; the supernatant is pipetted off to be used as cell-free extract immediately. Cell-free extracts should be stored on ice while waiting to be assayed. Short-term storage (on the order of a few days) at 4 °C usually does not negatively affect activity, although different enzymes will exhibit differing stability during storage. While cell-free extract can be frozen, it is not recommended as the freeze–thaw cycles can affect enzyme activity.

4.3 Expression and Purification of T. saccharolyticum Proteins in E. coli

4.3.1 Protein Expression Under Aerobic or Anaerobic Conditions

T. saccharolyticum proteins can be expressed and purified in *E. coli* following standard protein purification protocols. One commonly used strategy is to clone the *T. saccharolyticum* gene in the His-tag expression plasmid pEXP5-NT-TOPO [10]. Either anaerobic or aerobic conditions can be used to express *T. saccharolyticum* proteins in *E. coli*, depending on the oxygen sensitivity of the target protein. For anaerobic expression, the standard *E. coli* BL21 (DE3) expression protocol (manufacturer's protocol, New England Biolabs #C2527H) is followed until induction.

For anaerobic protein expression, an actively growing *E. coli* culture (i.e., an overnight culture) containing the expression plasmid is inoculated into 100 mL sterile LB broth with the appropriate antibiotic, and grown aerobically to OD_{600} ~0.5 with shaking at 200 RPM at 37 °C. The 100 mL *E. coli* culture is then transferred to a sterile serum bottle. In the case of T7-based expression plasmids, the appropriate amount of IPTG (Isopropyl β-D-1-thiogalactopyranoside) should be added to induce protein expression [10]. The serum bottle is then purged with 100% nitrogen to generate an anaerobic protein expression environment, and the culture is incubated for an additional 2–4 h at 37 °C before proceeding with harvesting.

4.3.2 Heat Treatment to Reduce Contamination from Native E. coli *Proteins*

E. coli cell-free extracts are prepared as described above, and then incubated at 50 °C anaerobically for 20 min, which denatures a portion of the native *E. coli* proteins (Fig. 8, lanes 2 and 3). The denatured proteins are separated with centrifugation at >12,000 × *g* for 5 min. The loss of *E. coli* proteins can be visualized on a Coomassie stained gel, as shown in Fig. 8 (purification of *T. saccharolyticum* AdhE protein is shown as an example).

To confirm that potential contamination from *E. coli* does not interfere with the target enzymatic activity, a control strain of *E. coli* should be used that contains an empty vector. After heat treatment, the cell extracts can be used for affinity column purification based on the type of tag used on the expression plasmid. Taking His-tagged *T. saccharolyticum* AdhE as an example, purification can be carried out using Ni-NTA spin columns (Qiagen #31314), according to the Qiagen protocol "Ni-NTA Agarose Purification of 6× His-tagged Proteins from *E. coli* under Native Conditions" with modifications as described previously [10]. The His-tagged AdhE can be eluted with 200 μL Elution Buffer (50 mM

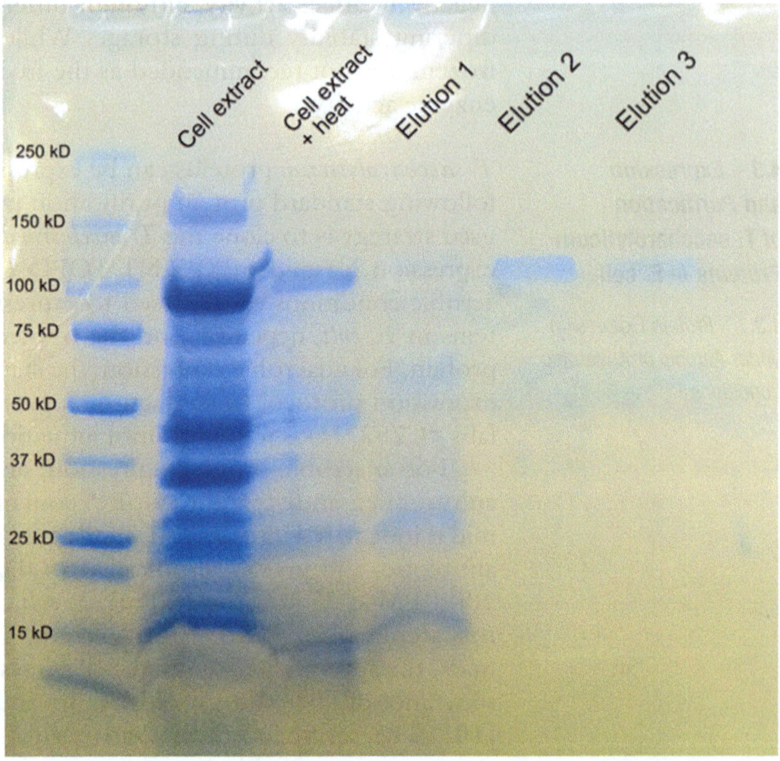

Fig. 8 Coomassie stained protein gel showing the effect of heat treatment and His-tag purification of *T. saccharolyticum* AdhE. Lane 1 shows the protein ladder; lanes 2 and 3 show the cell extracts before and after heat treatment; lanes 4 through 6 show the effect of three sequential elution fractions from the purification column: the target protein, AdhE, has a molecular weight of ~95 kD

NaH_2PO_4, 300 mM NaCl, 500 mM imidazole, 5 μM $FeSO_4$, pH 7), this is "Elution 1" in Fig. 7. Repeating this elution step sequentially generates more purified "Elution 2" and "Elution 3." For *C. thermocellum* and *T. saccharolyticum* AdhE, activity is measured at various stages during purification. The results of these activity measurements, along with Coomassie gels, can indicate the best elution fraction for usage in further enzymatic assays.

For best assay results, it is recommended that purified proteins be immediately used for enzymatic assays and stored on ice while waiting to be assayed. Some anaerobic proteins lose enzyme activity with freeze–thaw cycles [12, 22]. For longer-term storage of the purified protein, glycerol can be added to the sample buffer at a final concentration of 20% either via sample dialysis or with a PD-10 buffer changing column (GE Healthcare, Pittsburgh, PA). Enzymes stored in this manner should be checked for activity before use, since the ability to preserve activity by freezing varies from enzyme to enzyme.

4.4 Determining Relative Gene Expression

Relative gene expression in *T. saccharolyticum* is determined via quantitative real-time PCR (RT-qPCR), using the $2^{-\Delta\Delta CT}$ method. The expression of a gene of interest can be normalized against *recA* (*Tsac_1846*) expression [21]; *recA* is ideal as it has relatively stable expression at various growth phases.

Primers for RT-qPCR are designed with the online software, Primer3Plus (URL: http://www.bioinformatics.nl/cgi-bin/primer3plus/primer3plus.cgi); the primer length is generally kept to about 20 bp, and the primer melting temperature is kept at around 60 °C. The amplicon size should be between 100 and 120 bp. A convenient way to generate standards for RT-qPCR is to order a synthesized DNA template (i.e., ultramer or gBlock) that contains one copy of each of the amplicons of interest.

Acknowledgments

We thank Christopher Herring for useful discussion and critical review of the manuscript. We are grateful to Juergen Wiegel for providing us with a culture of *T. saccharolyticum* and advising us on genetic techniques. The manuscript has been authored by Dartmouth College under contract no. DE-AC05–00OR22725 with the U.S. Department of Energy. The BioEnergy Science Center is a U.S. Department of Energy Bioenergy Research Center supported by the Office of Biological and Environmental Research in the DOE Office of Science. Shuen Hon and Liang Tian have contributed equally to this work.

References

1. Liu SY, Gherardini FC, Matuschek M, Bahl H, Wiegel J (1996) Cloning, sequencing, and expression of the gene encoding a large S-layer-associated endoxylanase from Thermoanaerobacterium sp. strain JW/SL-YS 485 in *Escherichia coli*. J Bacteriol 178:1539–1547
2. Mai V, Lorenz WW, Wiegel J (1997) Transformation of Thermoanaerobacterium sp. strain JW/SL-YS485 with plasmid pIKM1 conferring kanamycin resistance. FEMS Microbiol Lett 148:163–167
3. Shaw AJ, Hogsett DA, Lynd LR (2010) Natural competence in Thermoanaerobacter and Thermoanaerobacterium species. Appl Environ Microbiol 76:4713–4719
4. Desai SG, Guerinot ML, Lynd LR (2004) Cloning of L-lactate dehydrogenase and elimination of lactic acid production via gene knockout in Thermoanaerobacterium saccharolyticum JW/SL-YS485. Appl Microbiol Biotechnol 65:600–605
5. Shaw AJ et al (2008) Metabolic engineering of a thermophilic bacterium to produce ethanol at high yield. Proc Natl Acad Sci U S A 105:13769–13774
6. Shaw AJ, Covalla SF, Hogsett DA, Herring CD (2011) Marker removal system for Thermoanaerobacterium saccharolyticum and development of a Markerless Ethanologen. Appl Environ Microbiol 77:2534–2536
7. Shaw AJ et al (2012) Urease expression in a Thermoanaerobacterium saccharolyticum ethanologen allows high titer ethanol production. Metab Eng 14:528–532
8. Herring CD et al (2016) Strain and bioprocess improvement of a thermophilic anaerobe for the production of ethanol from wood. Biotechnol Biofuels 9:125
9. Lo J, Zheng T, Hon S, Olson DG, Lynd LR (2015) The bifunctional alcohol and aldehyde dehydrogenase gene, adhE , is necessary for ethanol production in clostridium thermocellum and Thermoanaerobacterium saccharolyticum. J Bacteriol 197:1386–1393
10. Zheng T et al (2015) Cofactor specificity of the bifunctional alcohol and aldehyde dehydrogenase (AdhE) in wild-type and mutants of clostridium thermocellum and Thermoanaerobacterium saccharolyticum. J Bacteriol 197:2610–2619
11. Lo J et al (2015) Deletion of nfnAB in Thermoanaerobacterium saccharolyticum and its effect on metabolism. J Bacteriol 197: JB.00347-15
12. Zheng T et al (2017) Both adhE and a separate NADPH-dependent alcohol dehydrogenase gene, adhA; are necessary for high ethanol production in Thermoanaerobacterium saccharolyticum. J Bacteriol 199:e00542–e00516
13. Currie DH et al (2015) Genome-scale resources for Thermoanaerobacterium saccharolyticum. BMC Syst Biol 9:30
14. Janda JM, Abbott SL (2007) 16S rRNA gene sequencing for bacterial identification in the diagnostic laboratory: pluses, perils, and pitfalls. J Clin Microbiol 45:2761–2764
15. Olson DG, Lynd LR (2012) Transformation of clostridium thermocellum by electroporation. Methods Enzymol 510:317–330
16. Shaw AJ, Jenney FE Jr, Adams MWW, Lynd LR (2008) End-product pathways in the xylose fermenting bacterium, Thermoanaerobacterium saccharolyticum. Enzym Microb Technol 42:453–458
17. Tsakraklides V, Shaw AJ, Miller BB, Hogsett DA, Herring CD (2012) Carbon catabolite repression in Thermoanaerobacterium saccharolyticum. Biotechnol Biofuels 5:85
18. Hon S et al (2017) The ethanol pathway from Thermoanaerobacterium saccharolyticum improves ethanol production in clostridium thermocellum. Metab Eng 42:175–184
19. Herring CD et al (2012) Final report on development of Thermoanaerobacterium saccharolyticum for the conversion of lignocellulose to ethanol. United States Dep Energy, Washington, D.C. https://doi.org/10.2172/1033560
20. Bhandiwad A et al (2013) Metabolic engineering of Thermoanaerobacterium saccharolyticum for n-butanol production. Metab Eng 21:17–25
21. Zhou J et al (2015) Physiological roles of pyruvate ferredoxin oxidoreductase and pyruvate formate-lyase in Thermoanaerobacterium saccharolyticum JW/SL-YS485. Biotechnol Biofuels 8:138
22. Tian L et al (2016) Ferredoxin:NAD+ Oxidoreductase of Thermoanaerobacterium saccharolyticum and its role in ethanol formation. Appl Environ Microbiol 82:7134–7141
23. Caiazza N, Warner A, Herring C (2009) United States patent US2011/0059485 A1, Plasmids from thermophilic organisms, vectors derived therefrom, and uses thereof
24. Olson DG et al (2015) Identifying promoters for gene expression in clostridium thermocellum. Metab Eng Commun 2:23–29

25. Currie DH et al (2013) Functional heterologous expression of an engineered full length CipA from clostridium thermocellum in Thermoanaerobacterium saccharolyticum. Biotechnol Biofuels 6:32
26. Hogsett DAL (1995) Cellulose hydrolysis and fermentation by Clostridium thermocellum for the production of ethanol. (Dartmouth College)
27. Holwerda EK et al (2014) The exometabolome of clostridium thermocellum reveals overflow metabolism at high cellulose loading. Biotechnol Biofuels 7:155

Chapter 4

Methods for Metabolic Engineering of a Filamentous *Trichoderma reesei*

Yat-Chen Chou, Arjun Singh, Qi Xu, Michael E. Himmel, and Min Zhang

Abstract

In this work, we describe genetic tools and techniques for engineering *Trichoderma reesei* for the production of farnesene. Enhanced production of farnesene was used as an example of this methodology; as were the overexpression of a key enzyme, HMGS, in the MVA pathway.

Key words β-Farnesene, *Trichoderma reesei*, Direct microbial conversion

1 Introduction

Direct microbial conversion (DMC) offers a number of advantages for consolidating the processing units for biofuel production, thus increasing production efficiency and lowering overall process costs. Significant research advances have been made recently in the successful expression of cellobiohydrolases from ethanol-producing yeast [1]. Fermentation using the cellobiohydrolases and other cellulase-expressing yeast has shown that much lower enzyme loadings are required for fermenting solid biomass substrates (demonstrated by Mascoma Corporation). In addition, mild improvement of ethanol production from *Trichoderma reesei* was achieved by metabolic engineering strategies [2]. Efforts were made to enable an oligeaous yeast to express cellulases for the fatty acid production [3]. We have initiated work to extend the DMC concept to a "biomass to hydrocarbon production" scheme by first understanding the potential technical barriers and then devising strategies for improvement. To identify promising new host strains for engineering for hydrocarbon production from biomass feedstocks, two essential criteria were considered: (1) the strain should be either an excellent biomass degrader or (2) have extraordinary hydrocarbon-producing potential.

Michael E. Himmel and Yannick J. Bomble (eds.), *Metabolic Pathway Engineering*, Methods in Molecular Biology, vol. 2096, https://doi.org/10.1007/978-1-0716-0195-2_4,

Tricoderma reesei is an industrially important cellulolytic filamentous fungus [4, 5]. Because of *T. reesei*'s capacity to secrete large amounts of cellulases and hemi-cellulases, we reasoned that engineering this organism to produce hydrocarbon or its intermediates would generate an effective DMC strain of *T. reesei*. In general, engineering of secreting large amounts of cellulases has been difficult in fuel-producing organisms.

Here we have undertaken a project to introduce β-farnesene production into *T. reesei* and increase production by altering flux through the mevalonate pathway. To achieve this goal, we have developed and utilized advanced gene-targeting techniques and a markerless, recycleable selection strategy in the *T. reesei* system.

Aspects on the expression host strains, vector construction, and selection protocols for *T. reesei* have been discussed by Singh and coworkers [6]. Here we will continue the discussion and further provide several (successful) examples of the application of the mentioned aspects, focusing on the heterologous expression of farnesene production genes.

HMG-CoA synthase (HMGS) catalyzes the conversion of acetoacetyl CoA to HMG-CoA in MVP. Based on a study in *E. coli*, Kampranis et al. proposed up-regulating the yeast HMG-CoA synthase by a fusion enzyme strategy to increase the terpenoids production [7]. Furthermore, it was shown that overexpression of an equivalent gene to HMGS (of MVP pathway), *dxs*, encoding DXS (1-deoxy-D-xylulose-5-phosphate synthases in MEP pathway) improved the intracellular pool of isoprenoid precursors in *E. coli* [8]. Although *Trichoderma* uses MVP rather than MEP as in *E. coli*, overexpressing the first enzyme, HMG-CoA synthase (TrHMGS) in the isoprenoid pathway, may lead to an increase of the precursor pool for the end production (e.g., farnesene) formation.

2 Materials

2.1 Strains

T. reesei strain QM6a is the wild-type strain in this study. Strain 40–42 (*BF::112028, BF::53079, Δtmus53 Δpyr4*) was created by replacing each of the two nerolidol synthase genes with a β-farnesene synthase gene based on QM6a. AST1126 was created by random integration of a β-farnesene gene into QM6a.

2.2 Growth Media and Conditions

Potato dextrose (PD) was used as a routine medium to maintain *Trichoderma* strains. This medium contains 2% glucose. PDA (PD with agar) was used as a sporulation medium. For the selection of transformants, PDH plates (PD agar plates with 100 μg/mL hygromycin B) were used. The minimal Mandels Androtti Medium (MA) [9] was used as a base medium and modified as necessary. The pH of the medium was adjusted to 5.5 with KOH. Chemicals *trans*-β-farnesene (greater than 90% purity) and dodecane were purchased from Sigma-Aldrich (St. Louis, MO). All chemicals were used without further purification. All growth experiments were done at 30 °C in light. The liquid cultures were shaken at 225 rpm.

3 Methods

3.1 Plasmid Construction

3.1.1 Example: Construction of a Plasmid Expressing α- and β-Farnesene

The genes for α- and β-farnesene synthase, catalyzing the conversion of fanesyl pyrophosphate (FPP) to α- and β-farnesene (α-and β-FS) in *Picea abies* [10] and *Artemisia annua* [10], respectively, were codon optimized and synthesized for the expression in *T. reesei*. These genes were used to develop expression cassettes using *T. reesei enol* promoter and *T. reesei cbh1* transcription termination signal (Fig. 1, pTR63 as an example). *T. reesei* has the indigenous pathway to synthesize the intermediate FPP, which will now be channeled toward the synthesis of β-farnesene by the *T. reesei* transformants.

3.1.2 Construction of a Plasmid Overexpressing HMGS

To overexpress HMG-CoA synthase (HMGS) of *T. reesei*, we constructed a pUC19-based plasmid containing the following genes and DNA sequences (*see* Fig. 2):

1. Pgpd_HMGS_Ter CBH2 (*T. reesei* promoter for glyceraldehyde phosphate dehydrogenase driving the HMGS gene followed by a *T. reesei* terminator for CBH2).
2. Sh ble_pyr4_Sh ble (*T. reesei* pyr4 cassette flanked by Sh ble).

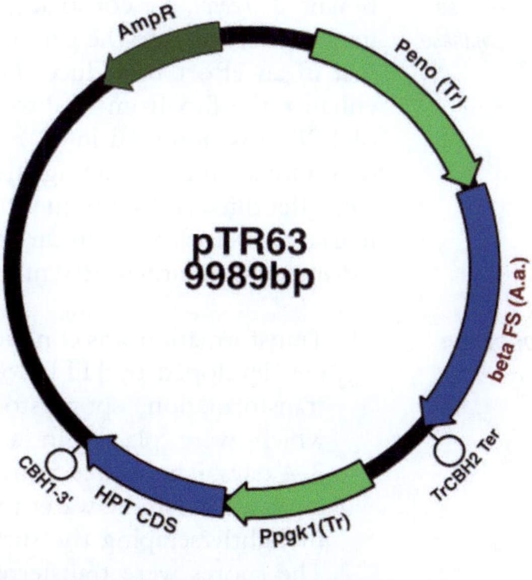

Fig. 1 Expression plasmid for β-farnesene synthase

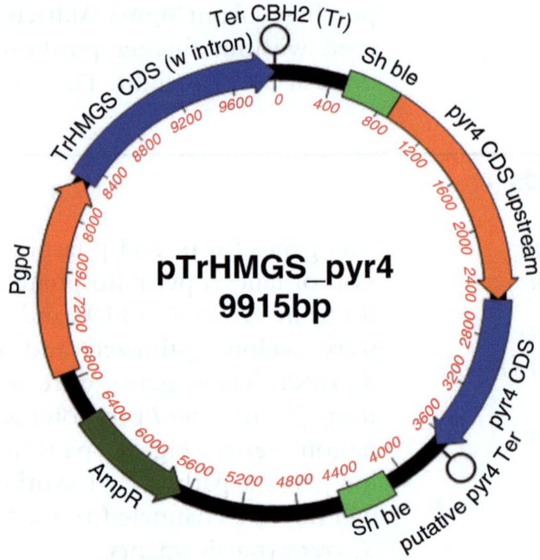

Fig. 2 Plasmid for the expression of HMGS in *T. reesei*

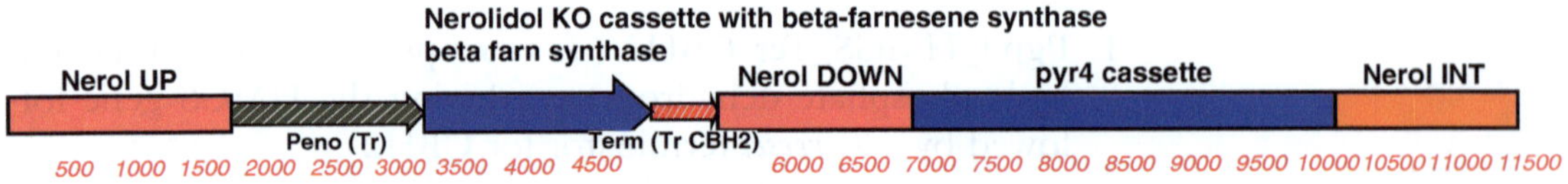

Fig. 3 Linear map of nerolidol synthase knockout cassette with β-farnesene synthase gene inserted

3.1.3 Construction of Nerolidol KO Cassettes with β-Farnesene Synthase Gene Insertion

In order to enhance the flux from FPP into β-farnesene in recombinant *T. reesei*, we constructed two β-farnesene cassettes for targeting insertion into the genome at two putative nerolidol synthase loci in an effort to reduce flux from FPP to nerolidol and also enhance the flux from FPP to β-farnesene. Nerolidol, a derivative of FPP, was observed in *T. reesei* cultures. By BLASTing known nerolidol synthase genes against the *T. reesei* genome, two targets were identified and subsequently knocked out with a recyclable *pyr4* marker cassette [6]. A linear map of nerolidol synthase knockout cassette with β-farnesene synthase gene inserted is shown in Fig. 3.

***3.2 Transformation of* T. reesei**

1. Transformation was conducted according to a modified protocol developed by [11] Goldman and coworkers. Lighted for transformation, spore stocks were spread onto PD plates, which were placed in a lighted incubator at 30 °C for 3–4 days. The spores from these plates were collected by adding 5 mL of sterile water to each plate and releasing the spores by lightly scraping the surface with a sterile culture spreader. The spores were transferred to Eppendorf tubes and centrifuged at 3000 × *g* at 4 °C for 3 min. The spore pellets were

washed twice with sterile cold 1 M sorbitol by suspending the pellets in the sorbitol and centrifuging them as described above. The washed spore pellets were resuspended in 1 mL sterile cold 1 M sorbitol.

2. For each transformation, 100 μL of spore suspension was transferred to a prechilled electroporation cuvette with 0.1 cm electrode gap. Linearized DNA was added at 0.5–2.0 μg in a total volume of 10 μL to the spores in the cuvette and mixed. For control transformation, 10 μL of sterile water was added to the spore suspension. The spores were then electroporated in a Bio-Rad Gene Pulser using the following settings: 1.8 kV, 800 Ω, and 25 μF. Immediately after delivering the pulse, 1.0 mL PD medium was added to each cuvette and mixed with the spores. Contents of each cuvette were transferred to a sterile test tube, which was then incubated at room temperature for 15 h.
3. A. For selection of transformants, 200 μL of the transformation mixture was spread on each PDH agar plate containing 100 μg/mL of hygromycin B. The plates were incubated at 30 °C in a lighted incubator for up to 5 days. The transformant colonies from these plates were transferred to PDH plates and incubated at 30 °C for 2 days. All transformants that grew well on these plates were transferred to other PDH plates for sporulation. After 3–5 days, spores from the transformants were streaked onto fresh PDH plates containing 0.1% Triton X-100. (*T. reesei* forms compact colonies on such plates). Spores from the isolated single colonies were again streaked on similar plates for a second round of purification. Spore preparations from purified transformants were collected by adding sterile water to the plates and releasing the spores by gently scraping the surface of the plates with sterile culture spreaders. The spore preparations were stored at 4 °C for up to 2 weeks. For long-term storage, spore suspensions were adjusted to 20% glycerol and stored at −80 °C.

   B. For *pyr4* recycling transformation, transformation spores were spread on MA+TritonX agar plates (without uridine) and incubated for 3 days at 30 °C. Single colonies were purified two more times by streaking on fresh MA+TritonX plates and incubation. After PCR confirmation, spores of the putative correct transformants are streaked onto MA+TritonX +uridine+5FOA for the loss of *pyr4* cassette.

3.3 Testing of Transformants for Farnesene Production

Baffled 125 mL shake flasks containing 25 mL of appropriate media were inoculated with ~10^7 spores. The flasks were incubated with shaking at 225 rpm at 28 °C or 30 °C. After 24 h of incubation, 5 mL of dodecane was added to each flask and incubation was continued. At intervals, 1 mL sample was taken from the dodecane phase of each culture and transferred into an Eppendorf tube. The samples were centrifuged at 14,000 rpm (18,407 × *g*) for 5 min to remove any water or biomass present, and 100 μL of the dodecane layer was transferred into HPLC vials for quantitation of farnesene by GC-MS analysis.

Acknowledgments

Funding was provided by the BioEnergy Science Center (BESC) and the Center for Bioenergy Innovation (CBI), from the U.S. Department of Energy Bioenergy Research Centers supported by the Office of Biological and Environmental Research in the DOE Office of Science. This work was authored in part by Alliance for Sustainable Energy, LLC, the Manager and Operator of the National Renewable Energy Laboratory for the U.S. Department of Energy (DOE) under Contract No. DE-AC36-08GO28308. The views expressed in the article do not necessarily represent the views of the DOE or the U.S. Government. The U.S. Government retains and the publisher, by accepting the article for publication, acknowledges that the U.S. Government retains a nonexclusive, paid-up, irrevocable, worldwide license to publish or reproduce the published form of this work, or allow others to do so, for U.S. Government purposes.

References

1. Ilmén M, den Haan R, Brevnova E, McBride J, Wiswall E, Froehlich A, Koivula A, Voutilainen SP, Siika-aho M, la Grange DC, Thorngren N, Ahlgren S, Mellon M, Deleault K, Rajgarhia V, van Zyl WH, Penttilä M (2011) High level secretion of cellobiohydrolases by *Saccharomyces cerevisiae*. Biotechnol Biofuels 4:30
2. Xu Q, Himmel ME, Singh A (2015) Production of ethanol from engineered Trichoderma reesei. In: Book chapter 11: direct microbial conversion of biomass to advanced biofuels. Elsevier, Amsterdam, p 197
3. Xu Q, Knoshaug EP, Wang W, Alahuhta M, Baker JO, Yang S, Wall TV, Decker SR, Himmel ME, Zhang M, Wei H (2017) Expression and secretion of fungal endoglucanase II and chimeri cellobiohydrolase I in the oleaginous yeast Lipomyces starkeyi. Microb Cell Factories 16:126
4. Peterson R, Nevalainen H (2012) *Trichoderma reesei* RUT-C30—thirty years of strain improvement. Microbiology 158:58
5. Seiboth B, Ivanova C, Seidl-Seiboth V (2011) *Trichoderma reesei:* a fungal enzyme producer for cellulosic biofuels. IntechOpen Limited, London
6. Singh A, Taylor LE II, Vander Wall TA, Linger J, Himmel ME, Podkaminer K, Adney WS, Decker SR (2015) Heterologous protein expression in Hypocrea jecorina: a historical perspective and new developments. Biotechnol Adv 33:142–154
7. Kampranis SC, Makris AM (2012) Developing a yeast cell factory for the production of terpenoids. Comput Struct Biotechnol J 3:1–7
8. Ajikumar PK, Tyo K, Carlsen S, Mucha O, Phon TH, Stephanopoulos G (2008) Terpenoids: opportunities for biosynthesis of natural product drugs using engineered microorganisms. Mol Pharm 5:167–190
9. Mandels M, Andreotti RE (1978) Problems and challenges in the cellulose to cellulose fermentation. Process Biochem 13:6
10. Neil R, McPhee S, Derek J (2008) Fuel compositions comprising farnesane and farnesane derivatives and method of making and using same: Google Patents
11. Goldman GH, Van Montagu M, Herrera-Estrella A (1990) Transformation of *TrichodermaHarzianum* by high-voltage electric pulse. Curr Genet 17:169–174

Chapter 5

Methods for Algal Protein Isolation and Proteome Analysis

Eric P. Knoshaug, Alida T. Gerritsen, Calvin A. Henard, and Michael T. Guarnieri

Abstract

Microalgae present promising feedstocks to produce renewable fuel and chemical intermediates, in part due to high storage carbon flux capacity to triacylglycerides or storage carbohydrates upon nutrient deprivation. However, the mechanism(s) governing deprivation-mediated carbon partitioning remain to be fully elucidated, limiting targeted strain engineering strategies in algal biocatalysts. Though genomic and transcriptomic analyses offer key insights into these mechanisms, active post-transcriptional regulatory mechanisms, ubiquitous in many microalgae, necessitate proteomic and post-translational (e.g., phospho- and nitrosoproteomic) analyses to more completely evaluate algal responsiveness to nutrient deprivation. Herein, we describe methods for isolating total algal protein and conducting proteomic, phosphoproteomic, and nitrosoproteomic analyses. We focus on methods deployed for the chlorophyte, *Chlorella vulgaris*, a model oleaginous alga with high flux to renewable fuel and chemical precursors.

Key words Microalgae, *Chlorella vulgaris*, Biofuels, Proteomics

1 Introduction

Microalgal biocatalysts present unique opportunities as microbial cell factories and feedstocks for the production of fuel and chemical intermediates [1–5]. These microbes offer several advantages relative to conventional industrial biocatalysts and terrestrial feedstocks, including photosynthetic cultivation potential on nonarable lands, in brackish and saline waters, and, due to their inherent capacity to convert atmospheric CO_2, no additional carbon requirements when grown photoautotrophically. Thus, microalgal biocatalysts display favorable characteristics for both economically viable and sustainable deployment for biofuel or biochemical production.

Microalgal metabolic plasticity presents the opportunity to leverage these microbes for photoautotrophic conversion to diverse product suites, ranging from fuel intermediates to nutraceuticals, and biopolymers. Within the context of transportation fuels, algal

Michael E. Himmel and Yannick J. Bomble (eds.), *Metabolic Pathway Engineering*, Methods in Molecular Biology, vol. 2096, https://doi.org/10.1007/978-1-0716-0195-2_5,

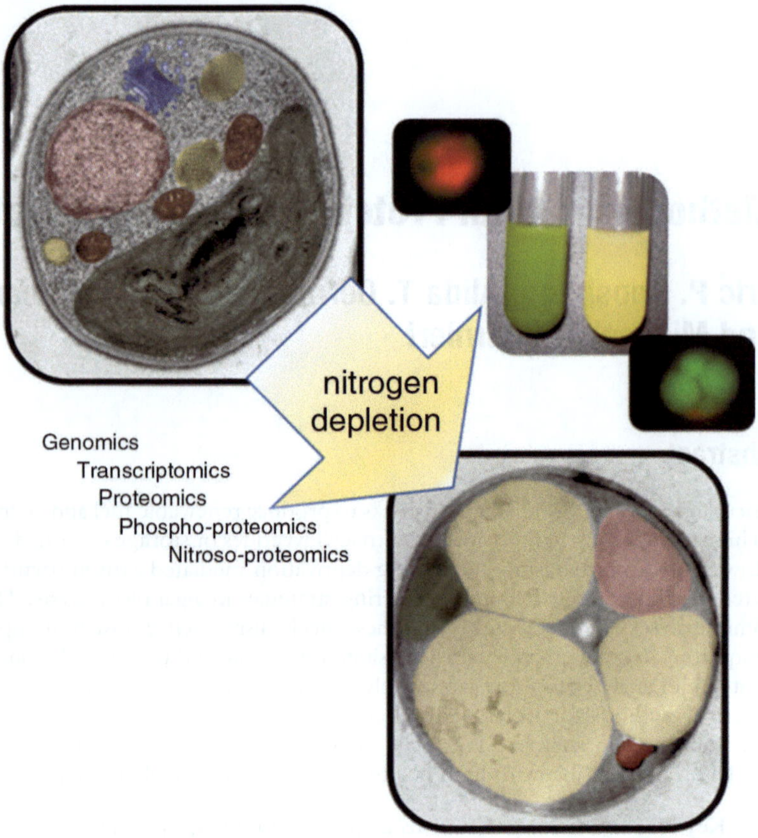

Fig. 1 Phenotypic changes in *Chlorella vulgaris* as the algae adjusts to nitrogen deplete conditions and the -omics studies that can lead to understanding this response. False color TEM images; Green highlight, chloroplast. Pink highlight, nucleus. Yellow highlight, oil bodies. Red highlight, mitochondria (TEM images courtesy of Bryon Donohoe, NREL). Culture tubes and fluorescent inset images (red fluorescence, chloroplast; green fluorescence, lipid bodies) show typical bulk culture coloration and BODIPY stained cells in nitrogen replete versus deplete conditions

flux to storage carbon in the form of triacylglyerides (TAG) is of primary consideration. TAG can be readily transesterified to fatty acid methyl esters (FAMEs) for use in renewable diesel blends. Oleaginous microalgae have been shown to accumulate high levels of intracellular lipid (>60% dcw) under physiological stress, such as nutrient deprivation (Fig. 1).

Nitrogen deprivation, in particular, has been extensively explored as a means to induce lipogenesis and study mechanisms thereof [6–8]. Such analyses have implicated post-transcriptional regulation as an integral factor in algal nutrient sensing and carbon partitioning, underscoring the value of proteomic and post-translational analyses in generating a complete algal metabolic pathway knowledge base. However, unlike most model microbial systems, algae present a series of unique challenges for total protein isolation and analysis, including robust cell walls and complex compartmentalization, hindering facile cell lysis and protein recovery and subsequent proteomic, post-translational modification, and computational analyses.

Here, we present a series of protocols for effective algal cell lysis, protein isolation, and proteomic analysis in the model oleaginous green alga, *Chlorella vulgaris*. Such methodology has enabled proteomic, phosphoproteomic, and nitrosoproteomic mapping of the alga [8–11], identifying key protein and protein-modifying regulators of TAG biosynthesis, and contributing to a comprehensive algal genome scale model [12].

2 Materials and Methods

2.1 Strains and Media

Chlorella vulgaris strain UTEX 395 was grown in 1 L Roux bottles using modified Bold's Basal Media (mBBM) containing: 2.94 mM $NaNO_3$, 0.17 mM $CaCl_2$, 0.30 mM $MgSO_4$, 0.43 mM K_2HPO_4, 1.00 mM KH_2PO_4, 0.43 mM NaCl, 0.17 mM EDTA, 18 μM $FeSO_4$, 0.18 mM H_3BO_3, 61 μM $ZnSO_4$, 15 μM $MnCl_2$, 10 μM MoO_3, 13 μM $CuSO_4$, and 3.3 μM $CoNO_3$. Cultures were maintained at 25 °C, with continuous (24 h) white fluorescent light illumination (200 μE m^{-2} s^{-1}). Cultures were supplemented with 2% CO_2 in air and mixed with a magnetic stir bar at 500 rpm. To induce nitrogen deprivation, rapidly growing cells were centrifuged for 5 min at 5000 × *g*, washed once in nitrogen-free mBBM, and resuspended in nitrogen-free mBBM for continued growth. Cell growth was monitored via cell count measurements using a 0.1 mm depth, Hy Lite_hemocytometer (Hausser Scientific). Cultures were inoculated at a cell density of approximately 3.5×10^6 cells mL^{-1}.

2.2 Algal Protein Isolation Protocol

1. Harvest 50 mL cell culture at cell densities of 7.85×10^7 cells mL^{-1} (nitrogen replete) and 5.00×10^8 cells mL^{-1} (nitrogen deplete) via centrifugation for 2 min at 3000 × *g*. Cultivation temperature (25 °C) is employed for centrifugation to ensure avoidance of cold-shock effects. Harvest densities were correlated to entry into log-phase growth for N-replete cells and will vary depending upon selected algal species.
2. Discard supernatants and quench cell pellets in liquid nitrogen. Pellets can be stored at −80 °C for extended periods (>2 weeks), though protein degradation was observed for *C. vulgaris* following 1 month of storage.
3. Thaw and solubilize pellets on ice in 2 mL of cold lysis buffer (50 mM Tris, pH 8.0, 150 mM NaCl, 1 mM DTT, 10% glycerol, supplemented with 1× cOmplete Protease Inhibitor

Cocktail solution (Roche Diagnostics Corporation, Indianapolis, IN). For phosphoproteomic analyses, Roche PhosStop protease inhibitor is also added to lysis buffer.

4. Sonicate cells on ice at 4 °C at 90% power setting for 30 s for 6–9 cycles, with a 1 min cool-down period between sonication cycles. A Braun-Sonic-L ultrasonicator was employed for all analyses. Sonication conditions will require fine-tuning for other algal strains and instrumentation. Microscopic visualization can aid in assessment of lysis efficiency.
5. Clear lysates via 2 cycles of centrifugation at 16,000 × *g* at 4 °C for 30 min.
6. Cleared lysates can be used directly in proteomic analyses or aliquoted and snap-frozen on liquid nitrogen for long-term storage at −80 °C.

2.3 Global Proteomic Analysis

To identify differential protein abundance under nitrogen replete and deplete conditions, gel-based liquid chromatography-mass spectrometry was employed for comparative global proteomic analysis as follows:

1. Separate 20 μg soluble proteins using one-dimensional SDS-PAGE.
2. Excise entire lanes from the gel and cut into 10–40 segments.
3. Reduce, alkylate, and tryptic digest gel segments robotically using a ProGest protein digestion station (DigiLab, Inc.) to generate peptide-containing liquid fractions suitable for LC/MS/MS analysis.
4. LC/MS/MS analysis was conducted on a Waters NanoAcquity HPLC system interfaced to a Thermo Fisher LTQ Orbitrap Velos mass spectrometer. Load peptides on a trapping column and elute over a 75 μm analytical column at 350 nL min^{-1}. Both columns were packed with Jupiter Proteo resin (Phenomenex).
5. Perform MS in the Orbitrap at 60,000 FWHM resolution operated in data-dependent mode. Select the 15 most abundant ions for LTQ MS/MS. For all proteomic analyses, a minimum of three biological replicates are required for downstream statistical analyses.
6. Search product ion data against the pertinent six-frame translational database appended with commonly observed background proteins to prevent false assignment of peptides derived from those proteins.
7. Parse Mascot DAT output files into the Scaffold program (Proteome Software) for validation and filtering to assess false discovery rates (FDR). Set scaffold parameters to a minimum of two peptides per protein with minimum probabilities of 95%

at the protein level and 50% at the corresponding peptide level (Prophet scores) in order to ensure <1% FDR.

8. Apply ANOVA statistical analysis and principal component analysis using ArrayTrack or suitable analysis software to identify differential significance between nutrient-replete and depleted samples, as well as between biological replicates (discussed further below). *Note*: only those positive protein identifications for which p-values less than or equal to 0.05 were obtained, were considered statistically significant for the data presented.
9. Apply data normalization based upon the total number of spectral counts under nitrogen-deplete conditions as described [13].

2.4 Phosphoproteomic Analyses

Algal cultures and lysates for phosphoproteomic analysis are generated as described above, with the following modifications:

1. Isolate cleared protein lysates as described in the Algal Protein Isolation protocol above. For phosphoproteomic analyses, Roche PhosStop protease inhibitor is added to lysis buffer in place of the cOmplete Protease Inhibitor Cocktail solution (Roche Diagnostics Corporation, Indianapolis, IN).
2. Digest 500 μg of lysate material in solution with the following concentration of buffer and enzymes: 100 mM DTT, 20 mM HEPES, 100 mM iodoacetamide, 25 ng μL^{-1} trypsin (final enzymes:substrate ratio of 1:20), and 20.5% trifluoroacetic acid (TFA).
3. Desalt 200 μg of peptide using solid phase extraction on a Waters C18 Sep Pak 1 cm^3 cartridge (part no. WAT054960) as follows: activate column with 1 mL 100% acetonitrile followed by 1 mL 50% acetonitrile and equilibrate with 3 × 1 mL 0.1% TFA.
4. Load sample on column and wash 2× with 1 mL 0.1% TFA followed by elution with 0.5 mL 60% acetonitrile and 0.1% TFA.
5. Dry elute to completion and enrich phosphopeptides using the GL Sciences TiO_2 kit (5010-21312) following manufacturer's instructions. Dry samples to completion and resuspend in 120 μL 0.1% TFA.
6. Analyze samples by nano LC/MS/MS with a Waters NanoAcquity HPLC system interfaced to a ThermoFisher Q Exactive.
7. Load peptides onto a trapping column and elute over a 75 μm analytical column at 350 nL min^{-1}. Columns are packed with Jupiter Proteo resin (Phenomenex) employing a 2 h gradient, as follows; buffer A: 0.1% formic acid in water; buffer B: 0.1% formic acid in acetonitrile; t_0 = 2%B, t_1 min = 5% B, t_{95}

min = 25% B, t_{110} min = 35% B, t_{112} min = 90% B, t_{113} min = 2% B, and t_{120} min = 2% B.

8. Operate mass spectrometer in data-dependent mode with MS performed in the Orbitrap at 70,000 FWHM resolutions and MS/MS performed using HCD and product ions detected in the Orbitrap at 17,500 FWHM resolutions. Select the 15 most abundant ions for MS/MS.
9. Process data through the MaxQuant software v1.5.1.0 (www.maxquant.org) for MS data recalibration using database filtering to 1% FDR, calculation of peak areas (label-free), and assignment of phosphosite localization probability.
10. Search data using a local copy of Andromeda with the following parameters: (1) enzyme: trypsin/P, (2) database: predicted proteins of algea species of interest with concatenated forward and reverse plus common contaminants, (3) fixed modification: carbamidomethyl (C), (4) variable modifications: oxidation (M), acetyl (protein N-term), phospho (STY), (5) fragment mass tolerance: 20 ppm.

2.5 Nitrosoproteomic Analyses

S-nitrosylated proteins are enriched from cleared protein lysates by using the biotin switch assay, as previously described [14]. Direct detection of an S-nitrosylated protein of interest can be detected via immunoblotting. Alternatively, the global nitrosomproteome can be evaluated by using mass spectrometry as described below:

1. Isolate cleared protein lysates as described in the Algal Protein Isolation protocol above and perform a biotin switch to enrich the S-nitrosylated proteins.
2. Purify and concentrate peptides after trypsin digestion using an online enrichment column (Thermo Scientific 5 μm, 100 μm ID × 2 cm C18 column).
3. Perform chromatographic separation on a reverse phase nanospray column (Thermo Scientific EASYnano-LC, 3 μm, 75 μm ID × 100 mm C18 column) using a 30 min linear gradient from 10% to 30% buffer B (100% acetonitrile, 0.1% formic acid) at a flow rate of 400 nL min^{-1}.
4. Elute peptides into the mass spectrometer (Thermo Scientific Orbitrap Velos) and collect spectra over a *m/z* range of 400–2000 Da using a dynamic exclusion limit of two MS/MS spectra of a given peptide mass for 30 s (exclusion duration of 90 s).
5. The instrument was operated in Orbitrap-LTQ mode where precursor measurements were acquired in the orbitrap (60,000 resolution) and MS/MS spectra (top 20) were acquired in the LTQ ion trap with a normalized collision energy of 35 kV. Compound lists of the resulting spectra were generated using

Xcalibur 2.2 software (Thermo Scientific) with an *S/N* threshold of 1.5 and 1 scan/group.

3 Proteomic Data Processing

Bioinformatics analysis of the phosphoproteomic data was performed by integrating differential expression analyses in the R language [15] with the UniProt (The Uniprot Consortium) database.

3.1 Differential Expression

1. Filter results by identified proteins from blastx search on NCBI. Remove all intensity values where there is no blastx hit and remove all rows where row sums are 0 (no intensity values for any samples).
2. Remove any rows where either two of three or all intensity measurements within a sample are 0 to reduce the chances of false positives. One sample condition may have zero readings in two of three replicates but the other may not (e.g., the depleted samples have intensity readings of 0, but the replete samples have readings), so the researcher must not remove the informative rows. This step was performed via conditional statements.
3. Analyze the remaining data via the edgeR package for differential expression [16]. Data must remain nonnormalized for the comparison to be relevant:
   (a) Read in the count table (intensities) and assign each column to a sample or group. In this case, there are two groups (depleted and replete) of three replicates each.
   (b) Filter rows with low intensity values below threshold.
   (c) Normalize readings across all samples (to account for low readings in some samples).
   (d) *Optional:* make an MDS plot to visualize the normalized data in two dimensions. If the data do not group by hypothesized biological separations or show groupings by replicate, the data should be examined for contamination or other errors.
   (e) Fit a negative binomial model to the data (see edgeR manual for explanation) and perform gene-specific dispersion estimate, allowing for variance between genes without assuming that genes across samples have the same mean-variance relationship.
   (f) Because this experiment was a simple comparison, an exact test was performed on each gene across the replete and depleted conditions. The test results were corrected

for FDR using a Benjamini-Hochberg correction and filtered for FDR ≧ 0.05. Results were written out to a tab-separated table for further analysis.

3.2 Proteomics Visualization

Visualizations were a qualitative assessment of the presence of proteins between the conditions and the putative functions of those proteins. Any proteins without functional annotation or gene ontology information were excluded from the visualizations. The following steps were performed:

1. Differential expression data tables were submitted as a tab-separated file to UniProt to search matching protein identifications and GO (Gene Ontology) numbers from the blastx identifiers.
2. Returned data were filtered for gene IDs, protein IDs, putative function, and GO terms and downloaded as a tab-separated file. This file was read into R as a table for annotation.
3. The differential expression results from **step 2** were merged with the UniProt results via the gene IDs to identify presence/absence and unique proteins between the conditions.
4. Frequencies of unique proteins and GO annotations were summarized across both conditions and sorted to determine which proteins were present only in one condition.
5. The most frequently occurring proteins in each condition were separated into their respective GO categories (molecular function, cellular component, biological process) and compared between the conditions via a bar chart.

Acknowledgments

This work was authored by Alliance for Sustainable Energy, LLC, the manager and operator of the National Renewable Energy Laboratory for the U.S. Department of Energy (DOE) under Contract No. DE-AC36-08GO28308. Funding provided by U.S. Department of Energy, Office of Energy Efficiency and Renewable Energy. Additional support was received from the Laboratory Directed Research and Development (LDRD) Program at the National Renewable Energy Laboratory. The views expressed in the article do not necessarily represent the views of the DOE or the U.S. Government. The U.S. Government retains and the publisher, by accepting the article for publication, acknowledges that the U.S. Government retains a nonexclusive, paid-up, irrevocable, worldwide license to publish or reproduce the published form of this work, or allow others to do so, for U.S. Government purposes.

References

1. Davis R, Markham J, Kinchin C, Grundl N, Tan ECD, Humbird D (2016) Process design and economics for the production of algal biomass: algal biomass production in open pond systems and processing through dewatering for downstream conversion. NREL/TP-5100-64772, Golden, CO
2. Dong T, Knoshaug EP, Davis R, Laurens LML, Van Wychen S, Pienkos PT, Nagle N (2016) Combined algal processing: a novel integrated biorefinery process to produce algal biofuels and bioproducts. Algal Res 19:316–323
3. EERE-BETO: Bioenergy Technologies Office National Algal Biofuels Technology Review (2016) https://www.energy.gov/sites/prod/files/2016/06/f33/national_algal_biofuels_technology_review.pdf
4. Knoshaug EP, Mohagheghi A, Nagle NJ, Stickel JJ, Dong T, Karp EM, Kruger JS, Brandner DG, Manker LP, Rorrer NA et al (2018) Demonstration of parallel algal processing: production of renewable diesel blendstock and a high-value chemical intermediate. Green Chem 20:457
5. Quinn JC, Davis R (2015) The potentials and challenges of algae based biofuels: a review of the techno-economic, life cycle, and resource assessment modeling. Bioresour Technol 184:444–452
6. Griffiths MJ, van Hille RP, Harrison STL (2014) The effect of nitrogen limitation on lipid productivity and cell composition in *Chlorella vulgaris*. Appl Microbiol Biotechnol 98:2345–2356
7. Guarnieri MT, Laurens LML, Knoshaug EP, Chou YC, Donohoe BS, Pienkos PT (2012) Complex systems engineering: a case study for an unsequenced microalga. In: Patnaik R (ed) Engineering complex phenotypes in Industiral strains. John Wiley & Sons, Inc., Hoboken, NJ, pp 201–231
8. Guarnieri MT, Nag A, Smolinski SL, Darzins A, Seibert M, Pienkos PT (2011) Examination of triacylglycerol biosynthetic pathways via de novo transcriptomic and proteomic analyses in an unsequenced microalga. PLoS One 6(10):e25851
9. Guarnieri MT, Gerritsen AT, Henard CA, Knoshaug EP (2018) Phosphoproteome of the oleaginous green alga, *Chlorella vulgaris* UTEX 395, under nitrogen-replete and -deplete conditions. Front Bioeng Biotechnol 6:19
10. Guarnieri MT, Nag A, Yang S, Pienkos PT (2013) Proteomic analysis of *Chlorella vulgaris*: potential targets for enhanced lipid accumulation. J Proteome 93:245–253
11. Henard CA, Guarnieri MT, Knoshaug EP (2017) The *Chlorella vulgaris* S-nitrosoproteome under nitrogen-replete and -deplete conditions. Front Bioeng Biotechnol 17:100
12. Zuñiga C, Li C-T, Huelsman T, Levering J, Zielinski DC, McConnell BO, Long CP, Knoshaug EP, Guarnieri MT, Antoniewicz MR et al (2016) Genome-scale metabolic model for the green alga *Chlorella vulgaris* UTEX 395 accurately predicts phenotypes under autotrophic, heterotrophic, and mixotrophic growth conditions. Plant Physiol 172:589–602
13. Zybailov B, Mosley AL, Sardiu ME, Coleman MK, Florens L, Washburn MP (2006) Statistical analysis of membrane proteome expression changes in *Saccharomyces cerevisiae*. J Proteome Res 5:2339–2347
14. Forrester MT, Foster MW, Benhar M, Stamler JS (2009) Detection of protein S-nitrosylation with the biotin-switch technique. Free Radic Biol Med 46:119–126
15. R Core Team. R: A language and environment for statistical computing. R Foundation for Statistical Computing, Vienna, Austria 2017. https://www.R-project.org
16. Robinson MD, McCarthy DJ (2010) Smyth GK: edgeR: a bioconductor package for differential expression analysis of digital gene expression data. Bioinformatics 26:139–140

Chapter 6

An Improved Leaf Protoplast System for Highly Efficient Transient Expression in Switchgrass (*Panicum virgatum* L.)

Chien-Yuan Lin, Hui Wei, Bryon S. Donohoe, Melvin P. Tucker, and Michael E. Himmel

Abstract

As a robust perennial C4-type monocot plant and a native species to North America, switchgrass (*Panicum virgatum*) has been evaluated and designated as a strong candidate bioenergy crop by the U.S. DOE. Although genetic modifications of switchgrass have been used to successfully reduce the recalcitrance of switchgrass biomass for biofuel production, the generation of transgenic switchgrass is still a slow and laborious process. A transient protoplast system can provide an excellent platform to accelerate the selection of genes-of-interest for tailoring switchgrass biomass. However, partially due to the lack of the complete genomic information, the attempts to optimize the transient protoplast system for switchgrass remain scarce. In this chapter, we provide an improved protocol for switchgrass protoplast isolation, increased transformation efficiency using CsCl gradient ultracentrifugation-derived plasmid DNA and extended application of the transient switchgrass protoplast system to analyze protein expression using western blot.

Key words Switchgrass, Protoplast, Transient expression system, GFP expression, CsCl gradient ultracentrifugation

1 Introduction

Cell walls are one of the distinguishing features of plant cells that exist as a barrier between the plasma membrane and external environment. Plant cell walls provide structural support of the plant cell, protection against chemical, biological and mechanical disruption, as well as sensing abiotic or biotic stimuli from the environment [1, 2]. The principal components of plant cell walls are cellulose, hemicelluloses, lignin, pectin and several structural proteins; while the composition of plant cell walls vary between primary cell walls (PCWs) and secondary cell walls (SCWs), and among cell, tissue and species types [3]. Removing plant cell walls using cell wall-digesting enzymes dissembles the cell wall polymers and releases wall-less plant cells known as protoplasts. Although cell wall is absent from the protoplasts, these isolated plant cells have

Michael E. Himmel and Yannick J. Bomble (eds.), *Metabolic Pathway Engineering*, Methods in Molecular Biology, vol. 2096, https://doi.org/10.1007/978-1-0716-0195-2_6,

been shown to retain most of their biological activities and physiological regulations. This attributes protoplasts a valuable transient model system for probing biological processes of plants [4, 5].

Another reason for the popularity of using plant protoplasts for plant biology research is that they can offer a unique differentiated cell system for studying cell-specific activities with high transfection efficiency but require low maintenance [4]. Since the first protoplast isolation using the root tips of tomato seedlings (*Lycopersicum esculentum) was* developed half a century ago [6], protoplast isolation protocols from various plants, such as *Arabidopsis* [7, 8], *Brassica* species [9, 10], carrot (*Daucus carota*) [11], cassava (*Manihot esculenta Crantz*) [12], rice [13, 14], maize (*Zea mays*) [4], tobacco [15], poplar [16, 17], grape berry (cabernet sauvignon) [18], lettuce (*Lactuca sativa*) [19], *Magnolia* species [20, 21], oil palm (*Elaeis guineensis*) [22], parsley (*Petroselinum crispum*) [23], and sugarcane [24] have been developed and optimized.

Much of the previous research on plant protoplast has focused on somatic fusion to improve plant breeding, especially for inducing novel variations or acquiring disease and herbicide resistance [25, 26]. Following advances in molecular biotechnology, the emphasis of protoplast system has shifted to a platform for rapid characterization of gene functions [27]. Furthermore, the development of versatile DNA delivery approaches, such as microinjection [28], electroporation [29], and PEG-mediated transfection [7] into protoplasts have shown the robustness of this transient system to explore plant biology and physiology, including cell wall regeneration [30], plant organogenesis [31], plant–pathogen interactions [23, 32], environmental stresses [33], protein subcellular localization [34], protein–protein interactions [35, 36], promoter activities [37, 38], gene regulatory networks [39], RNA regulation [40], and cell signaling pathways [14, 41].

Although the protoplast system is a transient assay, it has been shown to be a rapid and reliable tool to examine the biological and physiological processes. For example, it has been used to examine the stability of transgene expression in sugarcane [24] and gene silencing machinery in rice [42]. Moreover, transient expression system using plant protoplasts is useful for high-throughput development and validation of the latest plant biotechnologies, such as gene silencing [42–44] and targeted genome-editing technologies like the clustered regularly interspaced short palindromic repeats (CRISPR)–CRISPR-associated (Cas) system [45–50].

In the field of renewable energy, one of the efficient and effective methods for promoting biofuel production is to reduce the recalcitrance of lignocellulosic materials via genetic engineering. As a robust perennial C4-type monocot grass, switchgrass (*Panicum virgatum*) has been dedicated as an important energy crop by both U.S. DOE and USDA-sponsored research program [51, 52]. Switchgrass is also considered a valuable feedstock because it serves dual purposes as forage and as a bioenergy feedstock [53]. However, switchgrass is a self-incompatible and outcrossing monocot species, making it complex and time-consuming to modify using conventional breeding methods [54]. Successful efforts to improve the quality and quantity of switchgrass biomass were largely achieved using plant tissue culture approaches to manipulate the genetic and metabolic regulation of lignin biosynthesis [55–59].

The functional characterization of genes or promoters in switchgrass by producing transgenic switchgrass remains laborious processes [60, 61], but can be significantly accelerated by using transient gene expression assays in protoplasts [4]. The protoplast-based transient system is a promising solution to rapidly study the functionality of different genes or the transcriptional activity of different promoters [60]. Recently, this approach was successfully demonstrated using switchgrass protoplasts to implement targeted genome editing and obtain homozygous mutants, circumventing the difficulty of directly engineering switchgrass biomass due to its high ploidy level [62].

Partially due to the incomplete switchgrass genome dataset, until now, there are only two reports towards developing the switchgrass protoplast system [60, 61]. In 2008, Mazarei et al. reported the first protoplast isolation protocol from both switchgrass leaves and roots [60]. Almost a decade later, a faster, lower-cost protoplast isolation and transfection system using switchgrass leaf tissue was reported in 2016 that resulted in a transfection efficiency of 30% under optimized condition [61].

Recently, we have successfully developed a consolidated protocol for transgenic switchgrass production [63], which facilitates the production of transgenic switchgrass within 6 months. However, to facilitate the high-throughput gene or promoter selection for reverse genetic modification of switchgrass with better biomass quality and quantity, we pursued improving the transient switchgrass protoplast protocol. It has been suggested that protoplast transfection efficiency greater than 50% is important to obtain reliable and reproducible data from a protoplast system [7]. Based on the two previous switchgrass protoplast protocols [60, 61] and our own experience in improved-throughput protoplast protocol using *Populus trichocarpa* [17], here, we are providing an alternative method for switchgrass protoplast isolation using conventional cell wall-digesting enzymes (Cellulase Onozuka R-10 and Macerozyme R-10) and, importantly, a focus on improving the transfection efficiency using higher quality plasmid DNA for downstream molecular application.

In this chapter, we describe an improved protocol for switchgrass protoplast isolation and an alternative DNA extraction method to improve the transfection efficiency. Subsequently, with improved transfection efficiency, this protocol extends the system for transient protein expression analysis by western blot, which provides an additional platform for studying biological and physiological processes of switchgrass.

2 Materials

2.1 Chemicals

1. Plant cell-wall-digesting enzymes (Cellulase R-10 and Macerozyme R-10) were purchased from Yakult Pharmaceutical Industry Co., Ltd. (Tokyo, Japan).
2. All other chemicals for plasmid DNA purification and protoplast isolation/transfection were purchased from Sigma-Aldrich (St. Louis, MO).
3. All media are prepared using Milli-Q H_2O (18.2 MΩ·cm at 25 °C), sterilized by either autoclave or filtration (0.22 μm PES membrane), and stored at room temperature (RT) unless indicated otherwise (*see* **Note 1**).

2.2 Strains and Media

1. Switchgrass seed: *Panicum virgatum* cultivar Alamo (Ernst Conservation Seeds, Inc., Meadville, PA).
2. *Escherichia coli* Strain: TOP10 (C404010, Thermo Fisher Scientific, Waltham, MA).
3. Luria-Bertani (LB) medium: 10 g/L tryptone, 5 g/L yeast extract, 5/L g NaCl.
4. Terrific Broth (TB) medium: 24 g/L yeast extract, 12 g/L tryptone, 9.4 g/L dipotassium phosphate (K_2HPO_4), 2.2 g/L monopotassium phosphate (KH_2PO_4), 0.4% glycerol.

2.3 General Supplies and Equipment

1. Shaker.
2. Incubator.
3. High-pressure steam autoclave.
4. Direct-Q® water purification system (MilliporeSigma, Burlington, MA).
5. Sorvall™ ST 16 Centrifuge (Thermo Fisher Scientific, Waltham, MA).
6. Flask (250 mL, 4000 mL).
7. Falcon tube (15 mL, 50 mL).
8. 1.5 mL and 2 mL microcentrifuge tubes (Eppendorf, Hauppauge, NY).
9. 6-well clear polystyrene microplate (3516, Corning, Corning, NY).
10. Petri Dish (VWR, Radnor, PA).
11. Growth chamber (Percival Scientific, Perry, IA).
12. Standard plastic nursery pot (6 in.).

13. Metro-Mix 360 soil (SunGro Horticulture, Agawam, MA).
14. DU-800 spectrophotometer (Beckman Coulter, Brea, CA).
15. Polyallomer bell-top quick-seal centrifugation tube (Beckman Coulter, Brea, CA).
16. Optima XL-80K preparative ultracentrifuge (Beckman Coulter, Brea, CA).
17. Type 70 Ti fixed angle rotor (Beckman Coulter, Brea, CA).
18. 20-gauge needle (Becton Dickinson, Franklin Lakes, NJ).
19. 10 mL syringe (Becton Dickinson, Franklin Lakes, NJ).
20. Amicon ultra-15 centrifugal filter (MWCO 10 kDa; Sigma-Aldrich, St. Louis, MO).
21. Rapid-flow sterile disposable bottle top filter (Thermo Fisher Scientific, Waltham, MA).
22. 40 μm Fisherbrand® cell strainer (Thermo Fisher Scientific, Waltham, MA).
23. Hemocytometer (Hausse Scientific, Horsham, PA).
24. Cellometer Vision Trio (Nexcelom Bioscience, Lawrence, MA).
25. NanoDrop Spectrophotometer (Thermo Fisher Scientific, Waltham, MA).
26. Eclipse 90i digital microscopy system (Nikon, Melville, NY).
27. X-Cite®120 fluorescence illuminator (EXFO, Mississauga, Ontario) with a 120 W metal halide lamp, a 480/25 nm excitation filter, and a 530/25 nm emission filter.
28. iBlot® 2 gel transfer device (Thermo Fisher Scientific, Waltham, MA).

2.4 General Reagents and Solutions

1. Clorox® bleach.
2. Ethanol (70%, 95%).
3. 100% isopropanol.
4. Sterile Milli-Q H_2O.
5. Q5® hot start high-fidelity DNA polymerase (NEB, Ipswich, MA).
6. Solution I: 50 mM glucose, 25 mM Tris–HCl, 10 mM EDTA.
7. Solution II: 0.2 M NaOH, 1% SDS.
8. Solution III: 3 M Potassium acetate, 2 M acetic acid.
9. TE buffer: 10 mM Tris–HCl (pH 8.0), 1 mM EDTA.
10. 10 mM EDTA (pH 8.0).
11. Ethidium bromide: 10 mg/mL in H_2O (E1510, Sigma-Aldrich).
12. 3 M NaCl: 175.32 g/L NaCl dissolved in 1 L H_2O.

13. NaCl-saturated 1-butanol: Mix 20 mL of 3 M NaCl to 100 mL of butanol and wait overnight before use (butanol is at the upper phase after phase separation).
14. Enzyme digestion solution: 10 mM MES (pH 5.7), 600 mM mannitol, 1.5% (w/v) cellulase R-10 (Yakult), 0.75% Macerozyme R-10 (Yakult), 1 mM $CaCl_2$, 5 mM β-mercaptoethanol, 0.1% BSA [42].
15. W5 solution: 2 mM MES (pH 5.7), 125 mM NaCl, 125 mM $CaCl_2$, 5 mM KCl [42].
16. MMG solution: 4 mM MES (pH 5.7), 600 mM mannitol and 15 mM $MgCl_2$ [42].
17. Evans blue staining solution: 16 mg of Evans blue dye (E2129, Sigma, St. Louis, MO) in 40 mL MMG solution.
18. PEG solution: 40% polyethylene glycol (PEG)-4000 (81,240, Sigma-Aldrich, St. Louis, MO), 600 mM mannitol, 100 mM $CaCl_2$ [42].
19. WI solution: 4 mM MES (pH 5.7), 600 mM mannitol, 4 mM KCl [42].
20. NuPAGE® LDS sample buffer (NP0007, Thermo Fisher Scientific, Waltham, MA).
21. NuPAGE® reducing agent (NP0004, Thermo Fisher Scientific, Waltham, MA).
22. NuPAGE® Novex 4–12% bis-tris mini gel (NP0321BOX, Thermo Fisher Scientific, Waltham, MA).
23. iBlot® 2 transfer stack (IB24002, Thermo Fisher Scientific, Waltham, MA).
24. Monoclonal GFP tag antibody (11E5) (A-11121, Thermo Fisher Scientific, Waltham, MA).
25. Goat anti-mouse IgG, IgM (H+L) cross adsorbed secondary antibody, AP (31330, Thermo Fisher Scientific, Waltham, MA).
26. Chicken IgY polyclonal antibody against soybean ferritin from Lin et al. [64].

3 Methods

An efficient and reliable transient switchgrass protoplast isolation has not been well-established and has only been explored in two previously published studies [60, 61]. In the protocol presented here, we provide an alternative approach for switchgrass protoplast isolation, and importantly improve the transfection efficiency of the transient switchgrass protoplast system. The processes, including plant growth, protoplast isolation, plasmid DNA purification and PEG-transfection, are summarized in this protocol.

3.1 Plant Growth, Culture, and Conditions

The growth of switchgrass seeds was modified from Mazarei et al. [60].

1. Sterilize ~2 g of seeds with 70% ethanol in a 250 mL flask for 1 min.
2. Rinse with 50 mL of sterile water ten times.
3. Immerse the seeds in 100 mL sterile water and incubate overnight in the dark at 4 °C (*see* **Note 2**).
4. Transfer seeds to 6 in. pots (~65 seeds/180 cm^2) filled with soil (Metro-Mix 360) and grow in the green house.

The seeds following 70% ethanol sterilization and water rinse are shown in Fig. 1a. After water incubation (imbibition) and cold treatment (stratification), switchgrass seeds can be observed sprouting within 3–4 days of transferring the seeds to the pots. In parallel, we germinated dehusked seeds derived from our previous study [63], and the sprouted seeds were transferred to pots after water incubation for 4–5 days. Compared to the dehusked seeds, the imbibition and stratification approach appears to improve the rate and consistency of seedling growth (data not shown). Four-week-old switchgrass seedlings are about 12–13 cm tall and ready for protoplast isolation (Fig. 1e, left). By comparison, the dehusked seedlings were only 5–6 cm in height (Fig. 1e, right).

3.2 Protoplast Isolation

The ideal age of the switchgrass seedlings for protoplast isolation can vary from 14-day-old plants grown in greenhouse to 3-month-old plants regenerated from tissue culture [13, 60]. A recent report suggested the age of the switchgrass seedlings may not be a crucial factor for successful protoplast isolation. After the initial harvest, the authors showed that the established switchgrass lawns can be sustained up to 3.5 months without decreasing the yield of protoplasts [61]. Four-week-old switchgrass seedlings were chosen for this protocol, which provides sufficient plant materials for protoplast isolation.

3.2.1 Protoplast Isolation Procedure

1. Cut 4 g of the four-week-old switchgrass leaves using sharp single-edge blades into smaller leave strips (0.5–2 mm wide).
2. Immerse immediately the leave strips in 40 mL enzyme digestion solution in a 50 mL Falcon tube.
3. Incubate the leave strips in the enzyme solution for 3 h at RT with gentle rotation.
4. Transfer leave strips to a new 50 mL Falcon tube filled with 40 mL of W5 solution.

Fig. 1 Switchgrass seedling growth after seed sterilization. (**a**) Switchgrass seeds after sterilization. (**b**) 1-week-old seedlings. (**c**) 2-week-old seedlings. (**d**) 3-week-old seedlings. (**e**) 4-week-old seedlings. Scale bar = 2 cm

5. Shake gently for 5–10 min to release the switchgrass protoplasts.
6. Filter the protoplast suspension through a 40 μm cell strainer.
7. Centrifuge for 200 × *g* for 3 min and discard the supernatant carefully.
8. Add gently 10 mL of W5 solution in the Falcon tube to resuspend the protoplasts.
9. Centrifuge for 200 × *g* for 3 min and discard the supernatant carefully.
10. Add gently 10 mL MMG solution in the Falcon tube to resuspend the protoplasts.
11. Adjust the cell density using a haemocytometer to an average cell density of 2×10^5 cell/mL by adding MMG solution or resuspending the cells in a smaller volume of MMG solution.
12. Protoplast viability is evaluated by Evans blue staining assay described in Mazarei et al. [60] (*see* **Note 3**).

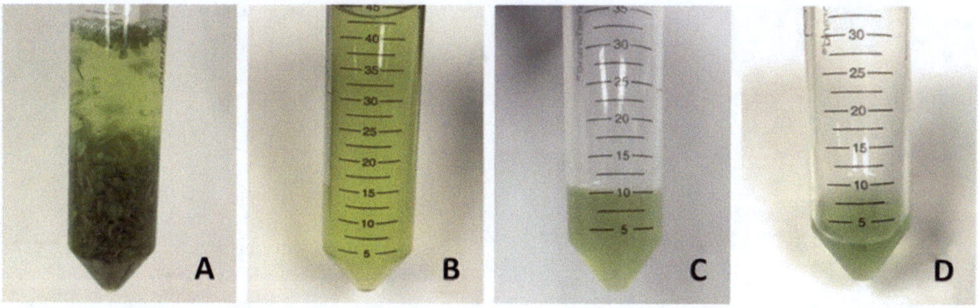

Fig. 2 The protoplast isolation procedure for switchgrass. (**a**) leave strips submerged in the digestion solution. (**b**) The filtrate of switchgrass protoplasts suspended in W5 solution. (**c**) Isolated protoplasts resuspended in 10 mL of MMG solution. (**d**) Isolated protoplasts resuspended in 5 mL of MMG solution to adjust the concentration

3.2.2 Protoplast Isolation Summary

After releasing the protoplast in W5, the clear, transparent W5 solution will become green and opaque, indicating the successful releasing of switchgrass protoplasts (Fig. 2a). The switchgrass protoplasts can be isolated by filtering the solution through a 40 μm nylon membrane to remove undigested leave strips and large cellular debris (Fig. 2b). Then, the cell density can be easily adjusted by resuspending of protoplasts in appropriate volume of MMG solution (Fig. 2c, d).

The yield of the switchgrass mesophyll protoplasts from 4 g of leaf strips as starting materials was ~3×10^6 cells (*see* **Note 4**). The released switchgrass protoplasts were observed by light microscopy and most of the protoplasts possess multiple chloroplasts due to their leaf tissue origin (Fig. 3a). The size of switchgrass protoplasts varied from 15 to 35 μm (Fig. 3b–d), which is similar to the cell population distribution measured using Cellometer with average cell diameter of 23.3 ± 0.6 μm. Protoplast viability was estimated using Evans blue staining to be 92.6 ± 3.6%, which is similar to the switchgrass protoplast viability (88–92%) from the leaf and root tissues of two switchgrass cultivars (Alamo and Alamo 2) reported in Mazarei et al. [60].

3.3 Plasmid Construction, Isolation, and Purification via Ultracentrifugation

We adopted the CsCl gradient method for plasmid DNA purification suggested by Yoo et al. [7] as an economical solution that provides larger DNA quantities and higher protoplast transfection efficiency (*see* **Note 5**). The CsCl gradient ultracentrifugation was modified according to Heilig et al. [65] and Lin et al. [17].

3.3.1 Plasmid Construction

To examine the efficiency of switchgrass protoplast transfection, we used fluorescence-based visualization of green fluorescence protein (GFP). The plasmid for transient GFP expression was modified based on pUC19 vector (NEB, Ipswich, MA) according to Chen et al. [66]. The pUC19 vector contains an ampicillin-resistant gene and a 54 base-pair multiple cloning site (MCS) for 13 different hexanucleotide-specific restriction endonucleases.

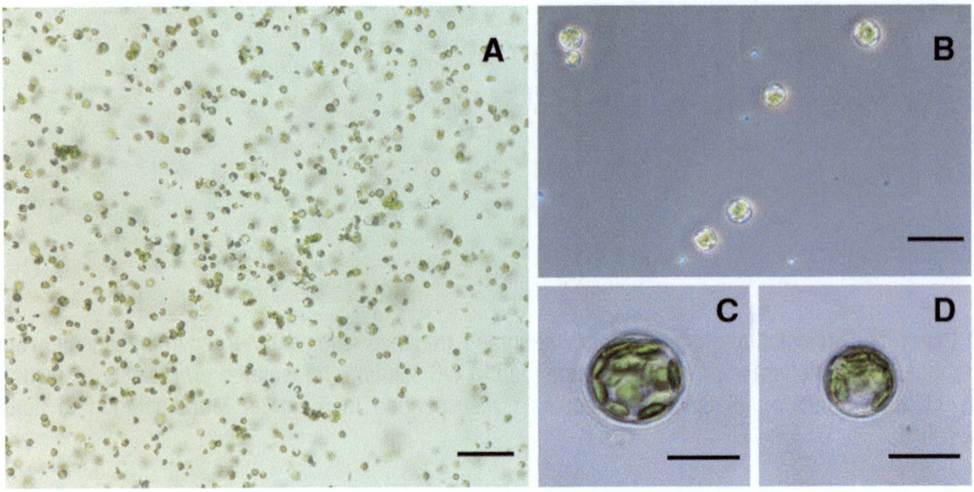

Fig. 3 Microscopic observation of the switchgrass protoplast population following isolation. (**a**) Switchgrass mesophyll protoplasts (Scale bar = 200 μm) (**b**) Switchgrass mesophyll protoplasts (Scale bar = 50 μm). (**c**) Larger switchgrass protoplast (Scale bar = 20 μm) (**d**) Smaller switchgrass protoplast (Scale bar = 20 μm)

Table 1
Primer set for subcloning GFP from pEGFP-N1 to pUC19-35S-Gus

Primer	Sequence (5′–3′)	Target	Amplicon (bp)
EGFP-F	ATCG**GGATCC**ATGGTGAGCAAGGGC	EGFP in pEGFP-N1	740
EGFP-R	CGAT**GAGCTC**TTACTTGTACAGCTC		

1. Excise the 35S promoter-driven β-glucuronidase (Gus) expression cassette (~3000 bp) from pBI121 using *Pst*I and *Eco*RI restriction enzymes and gel-purified before the ligation.
2. Ligate the Gus expression cassette into pUC19 using same enzyme set (*Pst*I and *Eco*RI) to construct pUC19-35S-Gus vector (~6000 bp).
3. Amplify the EGFP gene from pEGFP-N1 (Clontech, Mountain View, CA) using Q5 DNA polymerase and a specific primer set (EGFP-F and EGFP-R, Table 1).
4. Clone the EGFP gene fragment into the pUC19-35S-Gus using *BamH*I and *Sac*I to replace Gus and generate a pUC19-35S-EGFP (pGFP) vector (~4500 bp).
5. The pUC19-35S-eGFP-35S-ferritin (pGFP-Fer) vector (~6500 bp) was constructed by stacking the ferritin expression cassette from p1305.2-Ferritin [64] into pGFP using *Sph*I restriction enzyme.
6. Transform plasmid (pGFP or pGFP-Fer) into *E. coli* (TOP10).

3.3.2 Plasmid Isolation

1. Culture *E. coli* harboring the specific plasmid (pGFP or pGFP-Fer) in 3 mL LB media at 37 °C until OD_{600} reaches 0.6 and transfer 2 mL into 1000 mL of TB media in a 4000 mL flask [67] to grow for 12 h at 37 °C.
2. Harvest the cells by centrifugation at 6000 × *g* for 20 min.
3. Isolate the plasmid DNA by alkaline-lysis method [68, 69].
   (a) Resuspend *E. coli* pellet in 20 mL of solution I.
   (b) Lyse cells by adding 40 mL solution II and incubate at RT for 10 min.
   (c) Add 30 mL ice-cold solution III into the solution, mix gently and place on ice for 10 min.
   (d) Centrifuge the mixture at 10,000 × *g* for 30 min at 4 °C.
   (e) Transfer the supernatant to a new tube and precipitate the plasmid DNA by adding 0.6 volume of ice-cold 100% isopropanol to the supernatant.
   (f) Mix gently the solution and store at RT for 15 min.
   (g) Collect the precipitated plasmid DNA by centrifugation at 12,000 × *g* for 10 min.
   (h) Rinse the DNA pellet with 95% ethanol and air-dry.

3.3.3 Plasmid Purification via Ultracentrifugation

1. Dissolve the air-dried plasmid DNA in 19.8 mL of 10 mM EDTA.
2. Add 23.1 g CsCl and 1.23 mL of ethidium bromide (EtBr) in the DNA solution (*see* **Note 6**).
3. Heat seal the DNA-CsCl-EtBr solution into quick-seal centrifugation tubes.
4. Ultracentrifuge the well-sealed centrifugation tube at 371,000 × *g* at r_{max} for 48 h at RT using a fixed angle Ti70 rotor.
5. Recover the plasmid DNA by collecting the DNA band (lower) from the centrifugation tube using a syringe and needles.
6. After DNA recovery, fill the DNA-CsCl-EtBr solution using deionize water to 6 mL and remove EtBr from plasmid DNA repeatedly using same volume of NaCl-saturated 1-butanol until solution become transparent.
7. Desalt the DNA-CsCl solution (aqueous phase) by repeated Amicon Ultra-15 centrifugal filtration using 15 mL deionize water to remove the CsCl.
8. Adjust the plasmid DNA concentration to 2000 ng/μL.
9. Aliquot the plasmid DNA to 50 μL/tube and store at −80 °C.

3.4 Protoplast Transfection

PEG-mediated transfection is commonly used for switchgrass protoplast transfection [60, 61] and this method demonstrates high transfection efficiency in both dicot and monocot plants [7, 13]. We also adopted the PEG-mediated transfection in this protocol but the process was modified according to the protocols described in Yoo et al. [7] and Lin et al. [17]. In this protocol, two levels of transfection were conducted: (1) small-scale transfection for observation of GFP and calculation of transfection efficiency, and (2) large-scale transfection for protein expression analysis using western blot.

3.4.1 Small-Scale Protoplast Transfection for Fluorescent Observation

1. Add 10 μg of plasmid DNA in a volume of 10 μL in a 2 mL microcentrifuge tube.
2. Add 100 μL of protoplast solution (2×10^5 cell/mL in MMG solution) in the same tube and mix gently.
3. Add 110 μL of 40% PEG solution into the DNA–protoplast mixture.
4. Initiate the transfection by tapping the tube to homogenize the mixture.
5. Incubate the mixture at RT for 15 min.
6. Add 440 μL of W5 solution to the mixture and invert the tube gently several times to stop the transfection.
7. Collect the transfected protoplast by centrifugation at $200 \times g$ for 3 min.
8. Resuspend the protoplast in 1 mL of WI solution.
9. Transfer the 1 mL protoplast solution into a six-well microplate.
10. Add extra 1 mL of WI solution into the plate.
11. Incubate the protoplast solution in the dark at 25 °C for 12–24 h.
12. Transfer the transfected protoplasts into new 2 mL microcentrifuge tube.
13. Harvest the protoplasts by centrifugation at $200 \times g$ for 3 min.
14. Observe the green fluorescence of the GFP by fluorescence microscopy equipped with appropriate filter set.
15. The transfection efficiency is about 61.9 ± 3.5% ($n = 7$) (Fig. 4a, c), while no fluorescence signal was observed from the control pUC19 vector-transfected protoplast (Fig. 4b, d).
16. The protoplast transfection efficiency using different plasmid DNA amount (20 μg) was also evaluated. However, there was no significant difference in the transfection efficiency when 10 μg or 20 μg of plasmid DNA was used (Fig. 5) (*see* **Note 7**).

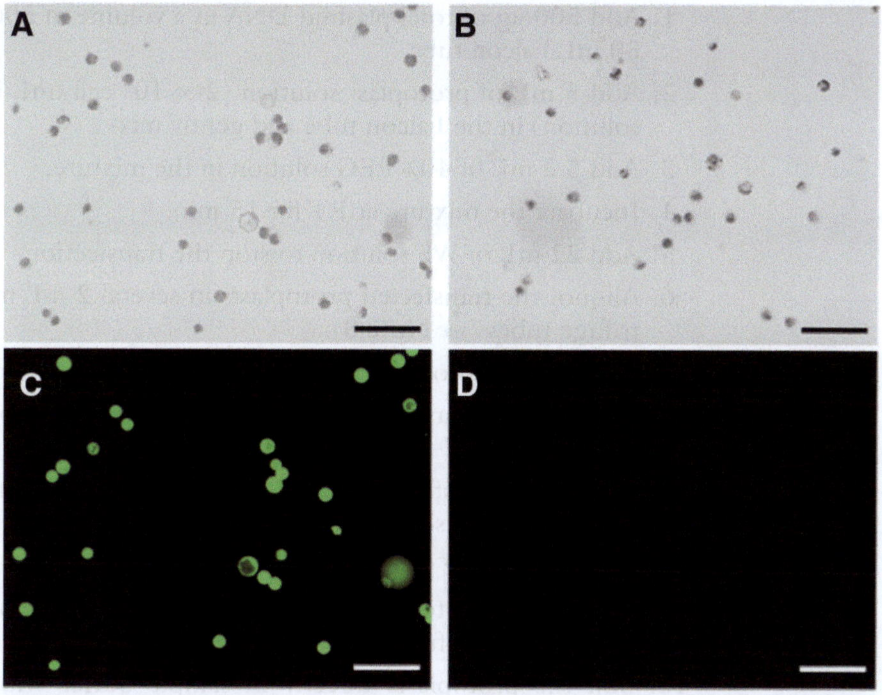

Fig. 4 Comparison of switchgrass protoplast the transfection efficiency between GFP containing vector (pGFP) and control vector (pUC19). (**a**) pGFP vector-transfected protoplast under bright field. (**b**) pUC19 vector-transfected protoplast under bright field. (**c**) Epi-fluorescence image of (**a**) under green fluorescence filter. (**d**) Epi-fluorescence image of (**b**) under green fluorescence filter. Scale bar = 200 μm

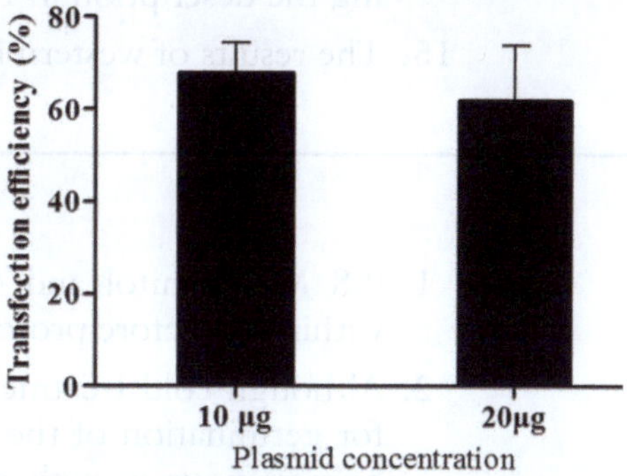

Fig. 5 Comparison of transfection efficiency when 10 or 20 μg plasmid DNA was used. Error bars represent standard error ($n = 5$)

3.4.2 Large-Scale Protoplast Transfection for western Blot

As suggested by Sheen (2001), the more effective PEG-mediated transfection method enables broader functional analyses of protein products of transgenes in protoplast transient assays [4]. We therefore extend the transient switchgrass protoplast assay to protein expression analysis by western blot.

1. Add 500 μg of total plasmid DNA in a volume of 500 μL in a 50 mL Falcon tube.
2. Add 5 mL of protoplast solution (2×10^5 cell/mL in MMG solution) in the Falcon tube and gently mix.
3. Add 5.5 mL of 40% PEG solution in the mixture.
4. Incubate the mixture at RT for 15 min.
5. Add 22 mL of W5 solution to stop the transfection.
6. Aliquot the transfected protoplasts in several 2 mL microcentrifuge tubes (*see* **Note 8**).
7. Collect the protoplasts by centrifugation at $200 \times g$ for 3 min.
8. Incubate the transfected protoplasts in 25 mL WI solution in Petri dish for 12–24 h.
9. Collect the transfected protoplasts by centrifugation at $200 \times g$ for 3 min using several 2 mL microcentrifuge tubes.
10. Combine all the protoplasts into a single tube.
11. Wash the collected protoplasts by 2 mL of W5 solution and collect by centrifugation at $200 \times g$ for 2 min.
12. Boil the protoplasts directly in sample buffer with sample reducing agent following manufacturer's protocol.
13. Separate the protein on a SDS-PAGE gel, and transfer to a polyvinylidene difluoride (PVDF) membrane using the iBlot 2 gel transfer device.
14. The general procedure of western blot was performed following the description in Lin et al. [64] (*see* **Note 9**).
15. The results of western blot are shown in Fig. 6.

4 Notes

1. 0.8 M mannitol and 40% PEG should be freshly prepared within 1 h before protoplast transfection.
2. Although cold treatment (stratification) may not be required for germination of the seeds, we experienced more rapid and homogenous growth of the switchgrass seedlings after cold treatment. Stratification is typically used to germinate *Arabidopsis* seeds to improve rate and synchrony of germination [70].
3. A similar protocol using Evans blue for protoplast viability determination can also be used [71].
4. Two conditions have been reported for switchgrass protoplast isolation using 16 healthy and young green leaves (4–5 cm long) [60] or >1.6 g of leaf tissue (14-day old) [61] with leave strips in size from 0.5–1 mm [60] to 2 mm [61]. Under

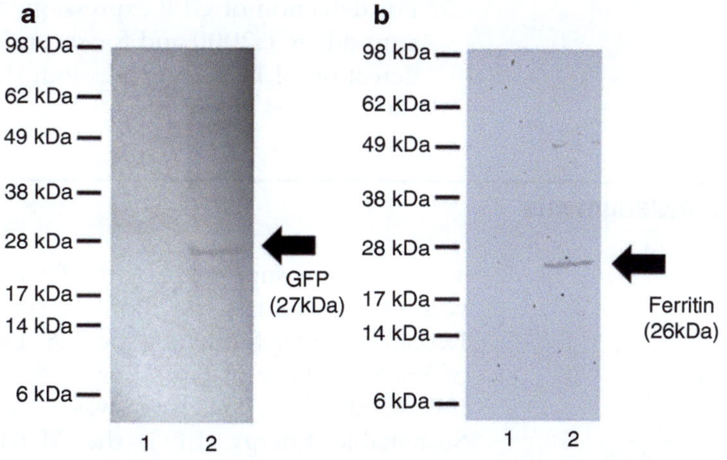

Fig. 6 Western blot analysis for transient protein expression. (**a**) pGFP vector-transfected protoplast. (**b**) pGFP-Fer vector-transfected protoplast. (Land 1) Control Vector (pUC19). (Lane 2) Plasmid DNA

the optimal conditions, ~7×10^5 protoplasts per isolation [60] and ~1.5×10^6 protoplasts per reaction [61] were achieved. Because we experienced proportionally higher yields of protoplast when starting with higher amounts of leaf tissue, we empirically determined that 4 g of leaf tissue as a starting material can consistently yield ~3×10^6 cells per isolation.

5. Plasmid DNA purification using CsCl/EtBr equilibrium centrifugation reproducibly yields a high-quality plasmid DNA free of most contaminants. Handling EtBr, a mutagen and potential carcinogen, requires care and caution.
6. At this step, some EtBr–protein complex precipitation will form after adding EtBr; therefore, a brief centrifugation at $1000 \times g$ for 10 min at RT can remove most of the precipitate and significantly benefit the plasmid DNA collection after ultracentrifugation [65].
7. Using a higher amount of plasmid DNA for protoplast transfection had been examined for improving transfection efficiency; however, when more than 40 μg plasmid DNA is used, it will likely precipitate after centrifugation. In addition, our transfection results were similar to Burris et al., who also showed no significant difference in transfection efficiency when 10 μg or 20 μg of plasmid DNA was used [61].
8. Aliquoting the 33 mL protoplast mixture into 2 mL before centrifugation is crucial to precipitate most of the protoplasts. Direct centrifugation of the entire 33 mL mixture will result in partial protoplast precipitation and loss of protoplasts.

9. For detection of GFP expression, the dilution for the primary antibody is 1:2000 and for secondary antibody is 1:2500. For detection of Ferritin expression, the antibody dilutions follow Lin et al. [64].

Acknowledgments

This work was supported by the Center for Direct Catalytic Conversion of Biomass to Biofuels (C3Bio), an Energy Frontier Research Center funded by the U.S. Department of Energy, Office of Science, Office of Basic Energy Sciences under Award Number DE-SC0000997. This work was authored in part by Alliance for Sustainable Energy, LLC, the Manager and Operator of the National Renewable Energy Laboratory for the U.S. Department of Energy (DOE) under Contract No. DE-AC36-08GO28308. The views expressed in the article do not necessarily represent the views of the DOE or the U.S. Government. The U.S. Government retains and the publisher, by accepting the article for publication, acknowledges that the U.S. Government retains a nonexclusive, paid-up, irrevocable, worldwide license to publish or reproduce the published form of this work, or allow others to do so, for U.S. Government purposes.

References

1. Ringli C (2010) Monitoring the outside: cell wall-sensing mechanisms. Plant Physiol 153:1445–1452
2. Humphrey TV, Bonetta DT, Goring DR (2007) Sentinels at the wall: cell wall receptors and sensors. New Phytol 176:7–21
3. Albersheim P, Darvill A, Roberts K, Sederoff R, Staehelin A (2010) Plant cell walls. Garland Science, New York
4. Sheen J (2001) Signal transduction in maize and arabidopsis mesophyll protoplasts. Plant Physiol 127:1466–1475
5. Li Q, Lin Y-C, Sun Y-H, Song J, Chen H, Zhang X-H, Sederoff RR, Chiang VL (2012) Splice variant of the SND1 transcription factor is a dominant negative of SND1 members and their regulation in *Populus trichocarpa*. Proc Natl Acad Sci U S A 109:14699–14704
6. Cocking EC (1960) A method for the isolation of plant protoplasts and vacuoles. Nature 187:962–963
7. Yoo S-D, Cho Y-H, Sheen J (2007) Arabidopsis mesophyll protoplasts: a versatile cell system for transient gene expression analysis. Nat Protoc 2:1565–1572
8. Wu F-H, Shen S-C, Lee L-Y, Lee S-H, Chan M-T, Lin C-S (2009) Tape-Arabidopsis Sandwich - a simpler Arabidopsis protoplast isolation method. Plant Methods 5:16
9. Hu Q, Andersen SB, Hansen LN (1999) Plant regeneration capacity of mesophyll protoplasts from Brassica napus and related species. Plant Cell Tissue Organ Cult 59:189–196
10. Uddin MJ, Robin AHK, Raffiand S, Afrin S (2015) Somatic embryo formation from co-cultivated protoplasts of Brassica rapa & *B. juncea*. Am J Exp Agric 8:342–349
11. Maćkowska K, Jarosz A, Grzebelus E (2014) Plant regeneration from leaf-derived protoplasts within the Daucus genus: effect of different conditions in alginate embedding and phytosulfokine application. Plant Cell Tissue Organ Cult 117:241–252
12. Wu J-Z, Liu Q, Geng X-S, Li K-M, Luo L-J, Liu J-P (2017) Highly efficient mesophyll protoplast isolation and PEG-mediated transient gene expression for rapid and large-scale gene characterization in cassava (*Manihot esculenta* Crantz). BMC Biotechnol 17:29
13. Zhang Y, Su J, Duan S, Ao Y, Dai J, Liu J, Wang P, Li Y, Liu B, Feng D et al (2011) A highly efficient rice green tissue protoplast system for transient gene expression and studying

light/chloroplast-related processes. Plant Methods 7:30

14. Cao J, Yao D, Lin F, Jiang M (2014) PEG-mediated transient gene expression and silencing system in maize mesophyll protoplasts: a valuable tool for signal transduction study in maize. Acta Physiol Plant 36:1271–1281
15. Fischer R, Hain R (1995) Tobacco protoplast transformation and use for functional analysis of newly isolated genes and gene constructs. Methods Cell Biol 50:401–410
16. Guo Y, Song X, Zhao S, Lv J, Lu M (2015) A transient gene expression system in Populus euphratica Oliv. Protoplasts prepared from suspension cultured cells. Acta Physiol Plant 37:160
17. Lin Y-C, Li W, Chen H, Li Q, Sun Y-H, Shi R, Lin C-Y, Wang JP, Chen H-C, Chuang L et al (2014) A simple improved-throughput xylem protoplast system for studying wood formation. Nat Protoc 9:2194–2205
18. Wang H, Wang W, Zhan J, Huang W, Xu H (2015) An efficient PEG-mediated transient gene expression system in grape protoplasts and its application in subcellular localization studies of flavonoids biosynthesis enzymes. Sci Hortic 191:82–89
19. Sasamoto H, Ashihara H (2014) Effect of nicotinic acid, nicotinamide and trigonelline on the proliferation of lettuce cells derived from protoplasts. Phytochem Lett 7:38–41
20. Shen Y, Meng D, McGrouther K, Zhang J, Cheng L (2017) Efficient isolation of Magnolia protoplasts and the application to subcellular localization of MdeHSF1. Plant Methods 13:44
21. Huo A, Chen Z, Wang P, Yang L, Wang G, Wang D, Liao S, Cheng T, Chen J, Shi J (2017) Establishment of transient gene expression systems in protoplasts from liriodendron hybrid mesophyll cells. PLoS One 12:e0172475
22. Masani MYA, Noll GA, Parveez GKA, Sambanthamurthi R, Prüfer D (2014) Efficient transformation of oil palm protoplasts by PEG-mediated transfection and DNA microinjection. PLoS One 9:e96831
23. Kanofsky K, Lehmeyer M, Schulze J, Hehl R (2016) Analysis of microbe-associated molecular pattern-responsive synthetic promoters with the parsley protoplast system. In: Hehl R (ed) Plant synthetic promoters: methods and protocols. Springer New York, New York, NY, pp 163–174
24. Gao S-J, Damaj MB, Park J-W, Beyene G, Buenrostro-Nava MT, Molina J, Wang X, Ciomperlik JJ, Manabayeva SA, Alvarado VY et al (2013) Enhanced transgene expression in sugarcane by co-expression of virus-encoded RNA silencing suppressors. PLoS One 8: e66046
25. Evans DA (1983) Agricultural applications of plant protoplast fusion. Bio/Technology 1:253
26. Kumar A, Cocking EC (1987) Protoplast fusion: a novel approach to organelle genetics in higher plants. Am J Bot 74:1289–1303
27. H-i J, Yan J, Zhai Z, Vatamaniuk OK (2015) Gene functional analysis using protoplast transient assays. In: Alonso JM, Stepanova AN (eds) Plant functional genomics: methods and protocols. Springer New York, New York, NY, pp 433–452
28. Neuhaus G, Spangenberg G (1990) Plant transformation by microinjection techniques. Physiol Plant 79:213–217
29. Bates GW (1999) Plant transformation via protoplast electroporation. In: Hall RD (ed) Plant cell culture protocols. Humana Press, Totowa, NJ, pp 359–366
30. Xu X, Xie G, He L, Zhang J, Xu X, Qian R, Liang G, Liu J-H (2013) Differences in oxidative stress, antioxidant systems, and microscopic analysis between regenerating callus-derived protoplasts and recalcitrant leaf mesophyll-derived protoplasts of *Citrus reticulata* Blanco. Plant Cell Tissue Organ Cult 114:161–169
31. Nenz E, Varotto S, Lucchin M, Parrini P (2000) An efficient and rapid procedure for plantlet regeneration from chicory mesophyll protoplasts. Plant Cell Tissue Organ Cult 62:85–88
32. Fraiture M, Zheng X, Brunner F (2014) An Arabidopsis and tomato mesophyll protoplast system for fast identification of early MAMP-triggered immunity-suppressing effectors. In: Birch P, Jones JT, JIB B (eds) Plant-pathogen interactions: methods and protocols. Humana Press, Totowa, NJ, pp 213–230
33. Confraria A, Baena-González E (2016) Using Arabidopsis protoplasts to study cellular responses to environmental stress. In: Duque P (ed) Environmental responses in plants: methods and protocols. Springer New York, New York, NY, pp 247–269
34. Miao Y, Jiang L (2007) Transient expression of fluorescent fusion proteins in protoplasts of suspension cultured cells. Nat Protoc 2:2348–2353
35. Li J-F, Bush J, Xiong Y, Li L, McCormack M (2011) Large-scale protein-protein interaction analysis in Arabidopsis mesophyll protoplasts

by split firefly luciferase complementation. PLoS One 6:e27364

36. Bracha-Drori K, Shichrur K, Katz A, Oliva M, Angelovici R, Yalovsky S, Ohad N (2004) Detection of protein–protein interactions in plants using bimolecular fluorescence complementation. Plant J 40:419–427
37. Stracke R, Thiedig K, Kuhlmann M, Weisshaar B (2016) Analyzing synthetic promoters using Arabidopsis potoplasts. In: Hehl R (ed) Plant synthetic promoters: methods and protocols. Springer New York, New York, NY, pp 67–81
38. Matsumoto TK, Keith LM, Cabos RYM, Suzuki JY, Gonsalves D, Thilmony R (2013) Screening promoters for Anthurium transformation using transient expression. Plant Cell Rep 32:443–451
39. Lee JH, Jin S, Kim SY, Kim W, Ahn JH (2017) A fast, efficient chromatin immunoprecipitation method for studying protein-DNA binding in Arabidopsis mesophyll protoplasts. Plant Methods 13:42
40. Zhang Z, Boonen K, Ferrari P, Schoofs L, Janssens E, van Noort V, Rolland F, Geuten K (2016) UV crosslinked mRNA-binding proteins captured from leaf mesophyll protoplasts. Plant Methods 12:42
41. Im JH, Yoo S-D (2014) Transient expression in Arabidopsis leaf mesophyll protoplast system for cell-based functional analysis of MAPK cascades signaling. In: Komis G, Šamaj J (eds) Plant MAP kinases: methods and protocols. Springer New York, New York, NY, pp 3–12
42. Bart R, Chern M, Park C-J, Bartley L, Ronald PC (2006) A novel system for gene silencing using siRNAs in rice leaf and stem-derived protoplasts. Plant Methods 2:13
43. Davey MR, Anthony P, Power JB, Lowe KC (2005) Plant protoplasts: status and biotechnological perspectives. Biotechnol Adv 23:131–171
44. Zhai Z, Sooksa-nguan T, Vatamaniuk OK (2009) Establishing RNA interference as a reverse-genetic approach for gene functional analysis in protoplasts. Plant Physiol 149:642–652
45. Nicolia A, Proux-Wéra E, Åhman I, Onkokesung N, Andersson M, Andreasson E, Zhu L-H (2015) Targeted gene mutation in tetraploid potato through transient TALEN expression in protoplasts. J Biotechnol 204:17–24
46. Zhang F, Maeder ML, Unger-Wallace E, Hoshaw JP, Reyon D, Christian M, Li X, Pierick CJ, Dobbs D, Peterson T et al (2010) High frequency targeted mutagenesis in *Arabidopsis thaliana* using zinc finger nucleases. Proc Natl Acad Sci U S A 107:12028–12033
47. Li JF, Aach J, Norville JE, McCormack M, Zhang D, Bush J, Church GM, Sheen J (2013) Multiplex and homologous recombination-mediated plant genome editing via guide RNA/Cas9. Nat Biotechnol 31:688–691
48. Jiang W, Zhou H, Bi H, Fromm M, Yang B, Weeks DP (2013) Demonstration of CRISPR/Cas9/sgRNA-mediated targeted gene modification in Arabidopsis, tobacco, sorghum and rice. Nucleic Acids Res 41:e188–e188
49. Woo JW, Kim J, Kwon SI, Corvalán C, Cho SW, Kim H, Kim S-G, Kim S-T, Choe S, Kim J-S (2015) DNA-free genome editing in plants with preassembled CRISPR-Cas9 ribonucleoproteins. Nat Biotechnol 33:1162
50. Li J-F, Zhang D, Sheen J (2015) Targeted plant genome editing via the CRISPR/Cas9 technology. In: Alonso JM, Stepanova AN (eds) Plant functional genomics: methods and protocols. Springer New York, New York, NY, pp 239–255
51. McLaughlin SB, Adams Kszos L (2005) Development of switchgrass (*Panicum virgatum*) as a bioenergy feedstock in the United States. Biomass Bioenergy 28:515–535
52. Perlack RD, Wright LL, Turhollow AF, Graham RL, Stokes BJ, Erbach DC (2005) Biomass as feedstock for a bioenergy and bioproducts industry: the technical feasibility of a billion-ton annual supply. National Lab TN, Oak Ridge
53. Guretzky JA, Biermacher JT, Cook BJ, Kering MK, Mosali J (2011) Switchgrass for forage and bioenergy: harvest and nitrogen rate effects on biomass yields and nutrient composition. Plant Soil 339:69–81
54. Xi Y, Fu C, Ge Y, Nandakumar R, Hisano H, Bouton J, Wang Z-Y (2009) Agrobacterium-mediated transformation of switchgrass and inheritance of the transgenes. Bioenergy Res 2:275–283
55. Fu C, Xiao X, Xi Y, Ge Y, Chen F, Bouton J, Dixon RA, Wang Z-Y (2011) Downregulation of cinnamyl alcohol dehydrogenase (CAD) leads to improved saccharification efficiency in switchgrass. Bioenergy Res 4:153–164
56. Shen H, He X, Poovaiah CR, Wuddineh WA, Ma J, Mann DGJ, Wang H, Jackson L, Tang Y, Neal Stewart C et al (2012) Functional characterization of the switchgrass (*Panicum virgatum*) R2R3-MYB transcription factor PvMYB4 for improvement of lignocellulosic feedstocks. New Phytol 193:121–136

57. Fu C, Mielenz JR, Xiao X, Ge Y, Hamilton CY, Rodriguez M, Chen F, Foston M, Ragauskas A, Bouton J et al (2011) Genetic manipulation of lignin reduces recalcitrance and improves ethanol production from switchgrass. Proc Natl Acad Sci U S A 108:3803–3808
58. Wuddineh WA, Mazarei M, Zhang J-Y, Turner GB, Sykes RW, Decker SR, Davis MF, Udvardi MK, Stewart CN (2016) Identification and overexpression of a knotted1-like transcription factor in switchgrass (*Panicum virgatum* L.) for lignocellulosic feedstock improvement. Front Plant Sci 7:520
59. Xu B, Escamilla-Treviño LL, Sathitsuksanoh N, Shen Z, Shen H, Percival Zhang YH, Dixon RA, Zhao B (2011) Silencing of 4-coumarate:coenzyme a ligase in switchgrass leads to reduced lignin content and improved fermentable sugar yields for biofuel production. New Phytol 192:611–625
60. Mazarei M, Al-Ahmad H, Rudis MR, Stewart CN (2008) Protoplast isolation and transient gene expression in switchgrass, *Panicum virgatum* L. Biotechnol J 3:354–359
61. Burris KP, Dlugosz EM, Collins AG, Stewart CN, Lenaghan SC (2016) Development of a rapid, low-cost protoplast transfection system for switchgrass (*Panicum virgatum* L.). Plant Cell Rep 35:693–704
62. Liu Y, Merrick P, Zhang Z, Ji C, Yang B (2018) Fei S-z: targeted mutagenesis in tetraploid switchgrass (*Panicum virgatum* L.) using CRISPR/Cas9. Plant Biotechnol J 16:381–393
63. Lin C-Y, Donohoe BS, Ahuja N, Garrity DM, Qu R, Tucker MP, Himmel ME, Wei H (2017) Evaluation of parameters affecting switchgrass tissue culture: toward a consolidated procedure for agrobacterium-mediated transformation of switchgrass (*Panicum virgatum*). Plant Methods 13:113
64. Lin C-Y, Jakes JE, Donohoe BS, Ciesielski PN, Yang H, Gleber S-C, Vogt S, Ding S-Y, Peer WA, Murphy AS et al (2016) Directed plant cell-wall accumulation of iron: embedding co-catalyst for efficient biomass conversion. Biotechnol Biofuels 9:225
65. Heilig JS, Elbing KL, Brent R (2001) Large-scale preparation of plasmid DNA. In: Current protocols in molecular biology. John Wiley & Sons, Inc., Hoboken, New Jersey
66. Chen H-C, Li Q, Shuford CM, Liu J, Muddiman DC, Sederoff RR, Chiang VL (2011) Membrane protein complexes catalyze both 4- and 3-hydroxylation of cinnamic acid derivatives in monolignol biosynthesis. Proc Natl Acad Sci U S A 108:21253–21258
67. Tartof K, Hobbs C (1987) Improved media for growing plasmid and cosmid clones. Focus 9:12
68. Sambrook J, Russell DW (2006) Preparation of plasmid DNA by alkaline Lysis with SDS: Maxipreparation. Cold Spring Harb Protoc 2006: pdb.prot4090
69. Nilsen TW (2013) RNA-friendly plasmid preparation. Cold Spring Harb Protoc 2013:pdb. prot072926
70. Bentsink L, Koornneef M (2008) Seed Dormancy and Germination. Arabidopsis Book 6: e0119
71. Danon A (2014) Protoplast preparation and determination of cell death. Bio-protocol 4: e1149

Chapter 7

Characterizing Intracellular Proteomes for Microbes: An Experimental Approach Using Label-Free Protein Quantitation

Paul E. Abraham and Robert L. Hettich

Abstract

Genomic and transcriptomic studies generate a rich source of molecular information; however, neither a static genome nor the presence of a messenger RNA provides the most direct insight into the functional state of a microbial cell at any given time. Rather, it is the ensemble of proteins (i.e., proteome) that is the primary workhorse for most biological processes that are concurrent and coordinately active in a living cell. Currently, mass-spectrometry-based (MS) technologies provide unprecedented information into the composition of the proteome, shedding light on the numerous complex biological processes actively shaping observed phenotypes. Herein, we detail a protocol for the exploration of intracellular proteomes through the large-scale, unbiased identification of proteins and their relative abundances using liquid chromatography (LC) coupled to tandem mass spectrometry (MS/MS). In general, this protocol is applicable to both microbial and eukaryotic systems.

Key words Proteomics, Label-free quantitation, LC-MS/MS

1 Introduction

Proteins carry out their functions at specific times and locations in the cell, sometimes with a physical or functional association to other proteins or biomolecules. The ensemble of protein molecules produced by a cell forms a highly connected biological network known as the proteome [1, 2]. Collectively, this network of proteins dynamically responds to external and internal perturbations to regulate the cell's functional state and define its phenotypes. Therefore, describing and understanding the quantitative composition of the proteome, as well as its dynamics, is a central and fundamental challenge of biology.

Be it a mechanistic, hypothesis-driven investigation or a large-scale, discovery-based investigation, the ability to identify the membership of the proteome is achieved through the utility of high-performance mass spectrometry (MS)-based technologies. Today,

Michael E. Himmel and Yannick J. Bomble (eds.), *Metabolic Pathway Engineering*, Methods in Molecular Biology, vol. 2096, https://doi.org/10.1007/978-1-0716-0195-2_7, © Springer Science+Business Media, LLC, part of Springer Nature 2020

the preferred method for quantifying the composition of a proteome or its constituents is bottom-up proteomics using data-dependent acquisition (DDA) [3]. In a bottom-up experiment, cells are lysed, and then the proteins are extracted and digested into peptides by a sequence-specific enzyme such as trypsin. The resulting proteolytic peptide mixtures are separated by liquid chromatography (LC) prior to the mass spectrometer, where they are measured and then fragmented to generate tandem mass spectra (MS/MS), which contain the information needed to identify, quantify, and link the specific peptides to their parent proteins. The resulting data is then analyzed by MS-specific computational pipelines.

Over the past several decades, MS-based proteomics has matured into a mainstream analytical tool for life sciences [4]. Herein, we describe a current protocol for bottom-up proteomics that is amenable to a wide variety of source materials, ranging from single microbial isolates to microbial communities in complex environment samples. It is a versatile method that supports the analysis of many aspects of protein identity, including sequence, quantity, and state of modification. Moreover, a major advantage of this protocol is that it is unbiased and free from hypotheses; that is, the researcher does not need to know the identity of the expected proteins in advance. Instead, this strategy relies on statistical associations to discover new biological relationships.

2 Materials

1. Ammonium bicarbonate (ABC) (Sigma-Aldrich, cat. no. 09830).
2. Sodium dodecyl sulfate (SDS) (Sigma-Aldrich, cat. no. L6026).
3. Sodium deoxycholate (SDC) (Sigma-Aldrich, cat. no. 30970).
4. Dithiothreitol (DTT) (Sigma-Aldrich, cat. no. 43815).
5. Iodoacetamide (IAA) (Sigma-Aldrich, cat. no. I1149).
6. Formic acid (FA) (Fischer, cat. No. A117).
7. Trifluoroacetic acid (TFA) (Sigma-Aldrich, cat. no. 302031).
8. LC-MS-grade water (Sigma-Aldrich, cat. no. WX0001).
9. LC-MS-grade acetonitrile (Sigma-Aldrich, cat. no. AX0156).
10. Ethyl acetate (Sigma-Aldrich, cat. no. 34972).
11. Chloroform (Sigma-Aldrich, cat. no. CX1054).
12. Eppendorf Safe-Lock microcentrifuge tubes (Sigma-Aldrich, cat. no. T9661).

13. Pipette tips (VWR, cat. Nos. 613-2295, 613-0299, 613-0344).
14. Sonicator (Branson 185 sonifier).
15. Bead beater (Bullet Blender 24 Storm).
16. Zirconium oxide beads (Next advance, cat. No ZrOB015 and ZrOB05).
17. Bench-top centrifuge (Eppendorf, cat. no. 5424 000.410).
18. Protein BCA concentration kit (Pierce, 23225).
19. Peptide concentration kit (Pierce, 23275).
20. UV spectrophotometer (Thermo Fischer, cat. no. 840208400).
21. C18 spin columns (Thermo Fischer, cat no. 89870).
22. Vivaspin 500 10 kDa MWCO (Sigma-Aldrich, cat. no. GE28-9322-25).
23. Thermomixer (Eppendorf Thermomixer Comfort, cat. no. 5382 000.015).
24. Mass spectrometer with high resolution and high mass accuracy; preferentially an Orbitrap-based instrument (Thermo Scientific Q-Exactive).
25. EASYspray ionization source for use with the PepMap columns (Thermo Scientific, cat. no. ES081), or Nanoflex source for use with alternative columns (Thermo Scientific).
26. Nanoflow HPLC system (EASYnLC 1000 from Thermo Scientific, cat. no. LC120).
27. MaxQuant data analysis software for peptide and protein identification and quantification (downloadable from http://www.maxquant.org/).
28. Crux data analysis software for peptide and protein identification and quantification (downloadable from http://crux.ms/).
29. Perseus data proteome analysis software (downloadable from http://www.biochem.mpg.de/5111810/perseus).
30. InfernoRDN proteome analysis software (downloadable from https://omics.pnl.gov/software/infernordn).
31. Vacuum concentrator (Thermo Scientific Savant SpeedVac, cat. no. SPD111VP1).
32. C18 spin columns (Sigma-Aldrich, cat. no. 66883-U).
33. Capillary columns (50 cm PepMap columns, 2 μm beads; Thermo Scientific Dionex, cat. no. ES803).
34. Dry ice.
35. −80 °C freezer.
36. Consumables (HPLC vials, caps, gloves, etc.).

37. Lysis buffer is composed of 2% (wt/vol) sodium dodecyl sulfate (SDS) in 100 mM ammonium bicarbonate (pH 8) and 10 mM dithiothreitol (DTT).
38. Denaturant buffer is composed of 2% (wt/vol) sodium deoxycholate (SDC) in 100 mM ammonium bicarbonate (pH 8).
39. HPLC solvents. Buffer A is 0.1% (vol/vol) formic acid (FA) in HPLC-grade water. Buffer B is 80% (vol/vol) acetonitrile in 0.1% (vol/vol) formic acid (FA) in HPLC-grade water.

3 Methods

3.1 Cell Lysis and Isolation of Proteins (Steps 1–4)

1. *Lysate preparation.* The starting amounts of cells or tissues affect the amount of protein that can be extracted and, if insufficient starting material is available, this will impact the depth of analysis. Unfortunately, because proteins can constitute a varying amount of the dry mass of a cell or tissue across organisms, there is not a standard recommended amount of starting biomass. Moreover, organisms will encode and produce a different number of protein molecules, so the amount of protein biomass needed for maximum coverage can also vary. In our current lab protocol, acceptable proteome coverage requires at least ~10 μg of extracted protein. Cells or tissues should be kept frozen prior to cell lysis, to avoid unwanted proteome alterations. To generate lysates from cells that are in suspension, it is necessary to pellet the cells by centrifugation (1000 × *g* for 10 min at 4 °C), discarding the supernatant, and then wash the pelleted cells with ice-cold PBS. To generate lysates from cells on culture dishes, place the cell culture dishes on ice, aspirate the medium and wash each dish once with ice-cold PBS. Scrape the cells in ice-cold PBS and transfer to a microcentrifuge tube. Pellet the cells by centrifugation (1000 × *g* for 10 min at 4 °C), discarding the supernatant. To generate lysates from tissue, freeze the tissue in liquid nitrogen and grind the tissue to a fine powder (*methodology pause point*: Cell pellets and ground tissue can be stored at −80 °C for several months).
2. *Cellular lysis.* As stated above, the amount of required starting mass will vary. The added lysis buffer should be kept at a minimum to avoid diluting the protein concentration. For maximum protein recovery, we recommend starting with a physical disruption using either bead beating or sonication. For bead beating, cells should be resuspended in an appropriate volume of lysis buffer before homogenization. For a microcentrifuge tube, the total volume of cells in the lysis buffer should be less than 300 μL to accommodate the required bead-to-sample proportion ratio (1-part bead and 2-parts sample). For

bead selection, use bead sizes that are approximately the same size as the sample you are trying to homogenize. We recommend 0.15 mm for most bacterial cells and 0.5 mm for eukaryotic cells. Instead of bead beating, sonication can be used for physical disruption of the cells. To sonically disrupt cells, suspend cells or tissue in a volume no less than ~250 μL. Program sonication to occur at an amplitude of 10% for 10 s on and 10 s off for 2 min. After completing the cell lysis step with or without bead beating or sonication, boil the samples in lysis buffer at 90 °C for 5 min. Centrifuge (12,000 × *g* for 10 min) to remove cellular debris and transfer supernatant to a new microcentrifuge tube.

3. *Block disulfide bond reformation*. Adjust samples to 30 mM IAA and incubate in the dark at room temperature to alkylate cysteine residues to prevent disulfide bond reformation. (*Important*: this step adds a static chemical modification of +57.02146 that needs to be included in the database search **step 12**.)

4. *Chloroform/methanol/water protein isolation*. Before performing this step, be aware of the additional volume requirements and make sure the sample tube has ample room to accommodate. Add ice-cold methanol to sample tube at a methanol-to-sample ratio of 4:1 and then vortex. Now add chloroform to sample tube at a chloroform-to-sample ratio of 1:1 and then vortex. Finally, add water to tube at a water-to-sample ratio of 3:1 and then vortex. Centrifuge (12,000 × *g* for 10 min at 4 °C) and discard the top and bottom layers. (*Note*: in our hands, this step is accomplished by carefully tilting the tube, allowing the protein precipitate to settle on the side of the microcentrifuge, and the two layers to empty.) Once isolated, wash the precipitated protein twice with cold methanol and then air dry the sample at room temperature.

3.2 Protein Digestion (Steps 5–10)

1. *Preparing proteins for enzymatic digestion*. Resuspend precipitated protein in denaturing buffer and incubate at 37 °C for 1 h.

2. *Protein concentration determination*. Perform a colorimetric protein concentration assay to determine the amount of protein in each sample.

3. *Protein digestion with protease*. Add sequencing-grade trypsin at a sample-to-protease ratio of 1:75 and digest overnight at 37 °C (*note*: for a single LC-MS/MS measurement in this protocol, only a maximum amount of ~5 μg of peptide mixture is required. Depending on the results from the protein concentration assay, if there is a surplus of protein material available, remove a standardized amount prior to digestion. For example, if 1 mg of protein is available for every sample, then remove

~100 μg of protein per sample for digestion). Add 100 mM ABC at twice the sample volume to dilute SDC to 1%. Add a second aliquot of trypsin at a sample-to-protease ratio of 1:75 and digest for 3 h at 37 °C.

4. *Removal of underdigested proteins.* Add sample on top of a 10 kDa molecular weight spin column filter and centrifuge (5000 × *g* for 10 min) to remove underdigested proteins. Collect the flow-through, which contains digested peptide mixture.
5. *Peptide purification and concentration.* Precipitate SDC from sample by acidifying the sample with TFA at a final concentration of 0.1% (vol/vol). Centrifuge (12,000 × *g* for 10 min) to pellet SDC and discard the pellet. Remove residual SDC by adding hydrated ethyl acetate to sample at twice the sample volume. Centrifuge (12,000 × *g* for 10 min) to liquid-phase extract the peptide mixture (i.e., bottom layer) (*note*: be careful to avoid aspirating the top or middle layer. Perform an additional ethyl acetate cleanup if needed). Vacuum dry the sample to dryness and resolubilize sample in 0.1% TFA. Peptide samples can be purified and concentrated from a variety of products. (*Note*: in our hands, the Thermo Scientific Pierce C18 Spin columns work best due to their flexible starting amounts and achieved peptide recovery.)
6. *Peptide concentration determination.* Perform a colorimetric peptide concentration assay to determine the amount of protein in each sample (*methodology pause point*: peptides can be stored at −80 °C for several months).

3.3 LC-MS/MS Analysis and Peptide Identification (Steps 11–13)

1. *Peptide sequencing.* Inject 1–5 μg of peptide mixture onto a preconditioned nano-LC column. Separate peptides across a 3 h linear gradient of 0–50% buffer B, followed by a 10 min wash with 90% buffer B, at a flow rate of 250 nL/min. Adjust the gradient and flow rates according to the column and LC system in use. Mass-spectrometric acquisition is performed by data-dependent acquisition, acquiring MS/MS spectra of the top 15 peptides from each MS scan. (*Note*: In any mass-spectrometric method, the number of identified proteins largely depends on the chromatographic and mass-spectrometric instruments and methods. In general, an intracellular proteome derived from a microorganism can be adequately profiled by a 3 h LC-MS/MS experiment. Higher coverage can be achieved by more sophisticated fractionation (e.g., 2D-LC-MS/MS) or LC equipment (e.g., UHPLC).)
2. *Perform data analysis with a freely available software.* At present, database searching algorithms remain the most frequently used method for large-scale protein identification. We

recommend using the MaxQuant software [5] or the Crux pipeline [6], as they are both widely used for peptide and protein identification as well as label-free quantification. From a simplified view, these algorithms perform the same basic functions—these programs take MS/MS spectra as input and score them against theoretical fragmentation patterns constructed from peptides generated from an in silico digestion of a user-supplied proteome database. The decoy database approach [7] is then used to ensure a suitable (usually 1%) false discovery rate (FDR) on the peptide and protein levels. While alternative software can be used, a low FDR is always necessary to obtain reliable data. For comparative proteome analysis, we recommend that all experiments are performed in triplicate and evaluated with standard statistical tests (*t*-test, Welch test, or ANOVA) to determine the significant quantitative changes. Freely available software, like InfernoRDN [8] and Perseus [9], can perform these tasks. One of the major "activation barriers" to large-scale data analyses is the numerous visualization and statistical procedures required to extract the maximum amount of information and biological insights. Once again, there are a variety of software tools that perform these tasks and generate graphic displays (e.g., heat-maps, volcano plots, PCA, correlation matrices) that are quite helpful for organizing the data into intuitive interfaces that allow researchers without formal computational skills to analyze their own data.

References

1. Zhu H, Bilgin M, Snyder M (2003) Proteomics. Annu Rev Biochem 72:783–812
2. de Hoog CL, Mann M (2004) Proteomics. Annu Rev Genomics Hum Genet 5:267–293
3. Aebersold R, Mann M (2016) Mass-spectrometric exploration of proteome structure and function. Nature 537(7620):347–355
4. Cravatt BF, Simon GM, Yates JR 3rd (2007) The biological impact of mass-spectrometry-based proteomics. Nature 450 (7172):991–1000
5. Cox J, Mann M (2008) MaxQuant enables high peptide identification rates, individualized p.p. b.-range mass accuracies and proteome-wide protein quantification. Nat Biotechnol 26 (12):1367–1372
6. McIlwain S et al (2014) Crux: rapid open source protein tandem mass spectrometry analysis. J Proteome Res 13(10):4488–4491
7. Elias JE, Gygi SP (2007) Target-decoy search strategy for increased confidence in large-scale protein identifications by mass spectrometry. Nat Methods 4(3):207–214
8. Polpitiya AD et al (2008) DAnTE: a statistical tool for quantitative analysis of -omics data. Bioinformatics 24(13):1556–1558
9. Tyanova S et al (2016) The Perseus computational platform for comprehensive analysis of (prote)omics data. Nat Methods 13(9):731–740

Chapter 8

Bacterial Differential Expression Analysis Methods

Sagar Utturkar, Asela Dassanayake, Shilpa Nagaraju, and Steven D. Brown

Abstract

RNA-Seq examines global gene expression to provide insights into cellular processes, and it can be particularly informative when comparing contrasting physiological states or strains. Although relatively routine in many laboratories, there are many steps involved in performing a transcriptomics experiment to ensure representative and high-quality results are generated for analysis. In this chapter, we present the application of widely used bioinformatic methodologies to assess, trim, and filter RNA-seq reads for quality using FastQC and Trim Galore, respectively. High-quality reads are mapped using Bowtie2 and differentially expressed genes across different groups were estimated using the DEseq2 R-Bioconductor package. In addition, we describe the various steps to perform the sample-wise data quality assessment by generating exploratory plots through the DESeq2 package. Simple steps to calculate the significant differentially expressed genes, up- and down-regulated genes, and exporting the data and images are also included. A Venn diagram is a useful method to compare the differentially expressed genes across various comparisons and steps to generate the Venn diagram from DESeq2 results are provided. Finally, the output from DESeq2 is compared to published results from EdgeR. The *Clostridium autoethanogenum* data are published and publicly available.

Key words Differential expression, *Clostridium autoethanogenum*, RNA-Seq, EdgeR, DESeq2

1 Introduction

RNA-Seq is an important technique used for a variety of applications that includes differential gene expression studies, identification of transcriptional features such as promoters, terminators, operons, and posttranscriptional control elements such as riboswitches and regulatory small RNAs (sRNAs) and it can detect weakly expressed genes [1, 2]. A variety of next-generation sequencing instruments can be used to generate data and different bioinformatic tools are used to analyze RNA-seq data sets, but the overall methodology across platforms is similar [3]. At present most expression studies collect cells, then extract and purify total RNA, followed by rRNA depletion or poly(A) enrichment [4]. Enriched transcript material is used as a template to generate complementary

Michael E. Himmel and Yannick J. Bomble (eds.), *Metabolic Pathway Engineering*, Methods in Molecular Biology, vol. 2096, https://doi.org/10.1007/978-1-0716-0195-2_8,

DNA (cDNA) libraries, via a reverse transcription enzymatic reaction, that then represents the pool of mRNA transcripts within each sample. DNA barcodes or adaptors are added to cDNAs during library creation, which permits different pooling of multiple samples that can be sequenced at the same time and so data can be assigned properly in subsequent analyses. Raw sequence reads are quality filtered/trimmed, typically aligned to a reference genome, then the number of reads mapped to each gene in the reference genome are counted and analyzed using a range of statistical and bioinformatic methods.

Key design of experiment considerations includes; ensuring high-quality RNA as the starting material, the number of biological replicates, the number of sequenced reads per sample, and read quality filtering and trimming [5]. A number of normalization methods have been developed and assessed to remove unwanted variance from RNA-seq data [6]. RNA-seq data sets often fit a negative binomial distribution [7–9] and a variety of software consider this during normalization. Other important considerations include the choice of mapping algorithm, its settings, the statistical test, and cost.

In this chapter, we apply bioinformatic methods to analyze a reference RNA-seq data set generated from *Clostridium autoethanogenum* growing under steady state using synthesis gas (syngas) at several gas–liquid mass transfer rates [10]. *C. autoethanogenum* is an acetogen capable of gas fermentation for fixation of carbon monoxide and carbon dioxide when hydrogen is present [11] and being commercialized deployed to utilize carbon in waste industrial off gasses from steel mills [12]. We apply highly cited bioinformatic methodologies to assess, trim, and filter RNA-seq reads for quality using FastQC and Trim Galore [13]. High-quality reads are mapped using Bowtie2 [14], reads counted with HTSeq [15] and differential expression is estimated for analysis using the DEseq2 software [16]. The output from DESeq2 is compared to earlier results from EdgeR using publicly available data [10].

RNA-Seq data are usually very large and require efficient algorithms that use minimum computing resources for their analysis. There are three main steps in the reference-based RNA-Seq analysis:

1. Aligning RNA-Seq reads to a reference genome or transcriptome.
2. Calculating the overlapping reads abundance (counts) against the gene/exon features.
3. Performing the differential expression analysis across different conditions.

Each of these steps includes several sub-steps, require bioinformatics expertise and possess different types of challenges. For each of these steps, different tools and algorithms are available, each offering their unique advantages. However, in this chapter, we only illustrate widely used algorithms published in many RNA-Seq analyses. The core programs used in our bioinformatics analysis include the following:

FastQC—This tool aims to provide a simple way to do some quality control checks on raw sequence data coming from high-throughput sequencing pipelines. It provides a modular set of analyses that can be used to get a quick impression of whether data have any problems that user should be aware of before performing further analysis.

Cutadapt—is an error-tolerant NGS-adapter trimming tool. It is primarily used by the Trim Galore script, described below. It also removes poly-A tails, primers and other unwanted sequences from sequencing data.

Bowtie2—is an ultrafast and memory-efficient tool for aligning sequencing reads to long reference sequences. Bowtie 2 indexes the genome with an FM Index to keep its memory footprint small and making the algorithm very efficient with large data. It also supports gapped, local, and paired-end alignment modes.

Trim Galore—is a wrapper script to automate quality and adapter trimming as well as quality control. For adapter trimming, Trim Galore uses the first 13 bp of Illumina standard adapters ("AGATCGGAAGAGC") as default. It allows specifying the Phred quality of base calls, and the stringency for adapter removal individually can remove the reads if they become shorter than specified length threshold during quality trimming and it also maintains the paired-reads status by removal of singletons. Optionally, it can automatically perform the FastQC on trimmed data with single parameter providing improved efficiency.

HTSeq—is a Python package providing infrastructure to process data from high-throughput sequencing assays. Although it can perform wide variety of sequence analysis tasks, the tool is most popular for generating the counts of aligned reads against the specified genomic features (i.e., genes).

DESeq2—is R-Bioconductor package to estimate variance-mean dependence in count data from high-throughput sequencing assays and test for differential expression based on a model using the negative binomial distribution.

2 Materials

2.1 Workspace Preparation

Before beginning the example run, it is recommended to create a working directory. Here, we have used a working directory called exp_run.

Note: In the following discussion executable commands are shown with a specific prefix.

1. Linux shell commands are shown with a "$" prefix.
2. Inline comments are shown with "#" prefix.
3. R interactive shell commands are shown with a ">" prefix.
4. R commands output are shown with a "##" prefix.
5. In the given path "/home/username," the <username> should be replaced with the current user's name as appropriate.

```
#Preparation of the workspace directory
$ cd /home/username
$ mkdir exp_run
$ cd exp_run
$ mkdir tools
```

2.2 Software and Tools

All software and tools utilized are open source and available under the GNU General Public License. All software was run using Linux command line and the DESeq2 program was run in an R interactive shell within RStudio. All installations were carried out in the `tools` directory and the precompiled executable binaries were added to the PATH environmental variable. Required R packages were installed using R-Bioconductor or *install.packages()* function. The software used require, for best results, a 64-bit version of the operating system, with at least 4 GB of RAM.

The basic installation commands for the essential tools are shown below.

2.2.1 SRA Toolkit

Download the latest version of SRA Toolkit for the Linux 64-bit architecture. Download the zip file from (https://www.ncbi.nlm.nih.gov/sra/docs/toolkitsoft/) and extract it.

```
#SRA Toolkit installation commands
$ cd /home/username/exp_run/tools/
$ tar -vxzf sratoolkit.tar.gz
$ export PATH = $HOME/exp_run/tools/sratoolkit:$PATH
```

2.2.2 FastQC

Download the latest version of the FastQC tool from (https://www.bioinformatics.babraham.ac.uk/projects/FastQC/). FastQC is a java application. To run it needs the system to have a suitable Java Runtime Environment (JRE) installed. Before you try to run FastQC, ensure that you have a suitable JRE. Installing FastQC is as simple as unzipping the zip file it comes in into a suitable location. Once unzipped it is ready to use.

```
#FastQC installation commands
$ cd /home/username/exp_run/tools/
$ unzip FastQC_v0.11.7
$ export PATH = $HOME/exp_run/tools/FastQC_v0.11.7:$PATH
```

2.2.3 Cutadapt

cutadapt is available from the Python Package Index (PyPI). To use cutadapt, you need Python 2.7 or 3.4 or above. Cutadapt follows install conventions of many Python packages. In the best case, it should install from PyPI as shown below:

```
#cutadapt installation commands
$ cd /home/username/exp_run/tools/
$ pip install cutadapt
```

2.2.4 Trim Galore

Download the latest version of Trim Galore from (https://www.bioinformatics.babraham.ac.uk/projects/trim_galore/). Trim Galore is a Perl wrapper around two tools: Cutadapt and FastQC. To use, ensure that these two pieces of software are available and copy the trim_galore script to a location available on the PATH.

```
#Trim Galore installation commands
$ cd /home/username/exp_run/tools/
# Check that cutadapt is installed
cutadapt --version
# Check that FastQC is installed
FastQC -v
# Install Trim Galore
curl -fsSL
https://github.com/FelixKrueger/TrimGalore/archive/0.4.5.tar.gz -o
trim_galore.tar.gz
tar xvzf trim_galore.tar.gz
$ export PATH = $HOME/exp_run/tools/TrimGalore-0.4.5:$PATH
```

2.2.5 Bowtie2 Software

Download Bowtie2 sources and binaries from the Download section of the Sourceforge site (http://sourceforge.net/projects/bowtie-bio/files/bowtie2). Binaries are available for the Intel x86_64 architecture running Linux, Mac OS X, and Windows. Download the binaries for Linux 64-bit system to the /home/username/exp_run/tools folder as a tarball and unpack them.

```
#Bowtie2 installation commands
$ cd /home/username/exp_run/tools/
$ unzip bowtie2-2.3.3.1-linux-x86_64.zip
$ cd bowtie2-2.3.3.1
$ export PATH = $HOME/exp_run/tools/bowtie2-2.3.3.1:$PATH
```

2.2.6 SAM Tools

For the SAM (Sequence Alignment Map) tools, download version 1.8 from (http://www.htslib.org/download/) as a tarball, and unpack.

```
#SAM Tools installation commands
$ cd /home/username/exp_run/tools/
$ tar -xvjf samtools-1.8.tar
$ cd samtools-1.8
$ ./configure --prefix=/home/username/exp_run/tools/
$ make
$ make install
$ export PATH = $HOME/exp_run/tools:$PATH
```

2.2.7 HTSeq

HTSeq is also available from the Python Package Index (PyPI) and can be installed like cutadpat.

```
#HTSeq installation commands
$ cd /home/username/exp_run/tools/
$ pip install HTSeq
```

2.2.8 DESeq2

DESeq2 was obtained from BioConductor, which requires pre-installation of R. R, Bioconductor, Rsubread, and DESeq2 were installed by following the below steps.

1. Download and install R and (https://www.r-project.org/) RStudio https://www.rstudio.com/products/rstudio/download2/).
2. Start the R program and install Bioconductor package by typing-

```
>source("https://bioconductor.org/biocLite.R")
>biocLite()
```

3. DESeq2 and its dependencies using.

```
>source("https://bioconductor.org/biocLite.R")
>biocLite("DESeq2")
```

4. For generating plots in R and DESeq2 requires additional libraries like data.table, ggplot2, pheatmap, RColorBrewer, gplots, and Vennerable, which can be installed in R directly by typing install.packages("package name) in R console. See the example below:

```
>install.packages("ggplot2")
```

All subsequent commands will assume that all tools are correctly installed and available within PATH variable.

2.3 RNA-Seq Data

In this example, we use a published study on *C. autoethanogenum* [10]. Raw RNA-seq data were obtained from the Sequence Read Archive (SRA) under the National Center for Biotechnology Information (NCBI) (http://www.ncbi.nlm.nih.gov/sra). The 12 data

Table 1
Sample attributes

SRA Accession	Physiological State
SRR5069221	Low biomass (0.5 gDCW/L) chemostat biological replicate 1
SRR5069222	Low biomass (0.5 gDCW/L) chemostat biological replicate 2
SRR5069223	Low biomass (0.5 gDCW/L) chemostat biological replicate 3
SRR5069224	Low biomass (0.5 gDCW/L) chemostat biological replicate 4
SRR5069225	Medium-low biomass (0.7 gDCW/L) chemostat biological replicate 1
SRR5069226	Medium-low biomass (0.7 gDCW/L) chemostat biological replicate 2
SRR5069227	Medium biomass (1.1 gDCW/L) chemostat biological replicate 1
SRR5069228	Medium biomass (1.1 gDCW/L) chemostat biological replicate 2
SRR5069229	Medium biomass (1.1 gDCW/L) chemostat biological replicate 3
SRR5069230	High biomass (1.4 gDCW/L) chemostat biological replicate 1
SRR5069231	High biomass (1.4 gDCW/L) chemostat biological replicate 2
SRR5069232	High biomass (1.4 gDCW/L) chemostat biological replicate 3

files are in the experiment with the series accession number corresponding to GSE90792 (*see* Table 1).

2.3.1 Sample Information

For demonstration, we have shown commands only for the first four samples. However, the same commands can be run on remaining samples by changing the respective file names.

2.3.2 Download SRA Data

The data were downloaded in the SRA format and converted to FASTQ using SRA Toolkit. The precompiled toolkit package is downloaded, and the locations of executables are added to the PATH or the bin folder of the current user. Then the following command is executed to download and convert the zipped SRA file to FASTQ:

```
#Download the SRA data from the NCBI SRA FTP site
$ cd /home/username/exp_run
$ mkdir raw_data
$ cd raw_data

$ wget ftp://ftp.ncbi.nlm.nih.gov/sra/sra-
instant/reads/ByRun/sra/SRR/SRR506/SRR5069221/SRR5069221.sra
$ wget ftp://ftp.ncbi.nlm.nih.gov/sra/sra-
instant/reads/ByRun/sra/SRR/SRR506/SRR5069222/SRR5069222.sra
$ wget ftp://ftp.ncbi.nlm.nih.gov/sra/sra-
instant/reads/ByRun/sra/SRR/SRR506/SRR5069223/SRR5069223.sra
$ wget ftp://ftp.ncbi.nlm.nih.gov/sra/sra-
instant/reads/ByRun/sra/SRR/SRR506/SRR5069224/SRR5069224.sra
```

Note: The SRA download links may have been updated. If links are no longer active, appropriate links can be searched through NCBI-SRA website (https://www.ncbi.nlm.nih.gov/sra) by providing the respective SRA accession numbers.

```
#Demultiplex "*.sra" files to FASTQ format with sra-toolkit
$ cd /home/username/exp_run/raw_data

$ fastq-dump -I --split-3  ./SRR5069221.sra
$ fastq-dump -I --split-3  ./SRR5069222.sra
$ fastq-dump -I --split-3  ./SRR5069223.sra
$ fastq-dump -I --split-3  ./SRR5069224.sra
```

2.4 Reference Genome Preparation

For the most RNA-Seq workflows, it is necessary to have the organism's reference genome sequence in FASTA format and annotation in suitable (GFF or GTF) format. The genome sequence is used to align the raw RNA-seq reads and GTF annotations are used to fetch the locations of the gene features. The easiest way to download the appropriate microbial reference genome sequence and annotation files is through Ensembl Bacteria browser (http://bacteria.ensembl.org/index.html), which houses over 90,000 prokaryotic genome assemblies, including multiple strains of many species. A representation of downloading *C. autoethanogenum* reference genome sequence in FASTA format and annotation in GTF format is shown (Fig. 1).

We downloaded the *C. autoethanogenum* reference genome sequence and GTF formatted annotations as shown above, unzipped the files and saved in the `reference_genome` directory. The FASTA and GTF files were renamed as `genome_ref.fasta` and `genome_ref.gtf` for convenience.

```
#Reference Genome Preparation
$ cd /home/username/exp_run
$ mkdir reference_genome

$ mv
Clostridium_autoethanogenum_dsm_10061.ASM48450v1.dna.chromo
some.Chromosome.fa ./reference_genome/genome_ref.gtf
$ mv Clostridium_autoethanogenum_dsm_10061.ASM48450v1.39.gtf
./reference_genome/genome_ref.gtf
```

Note: The name of the reference FASTA and GTF files may appear different than shown here.

Bowtie2 or for that matter most RNA-Seq aligner tools require reference genome to be indexed to fasten the alignment process before it is used. Genome indexing can be easily performed using the standard command available through Bowtie2 package. After indexing, the output contains a set of new files with user-specified

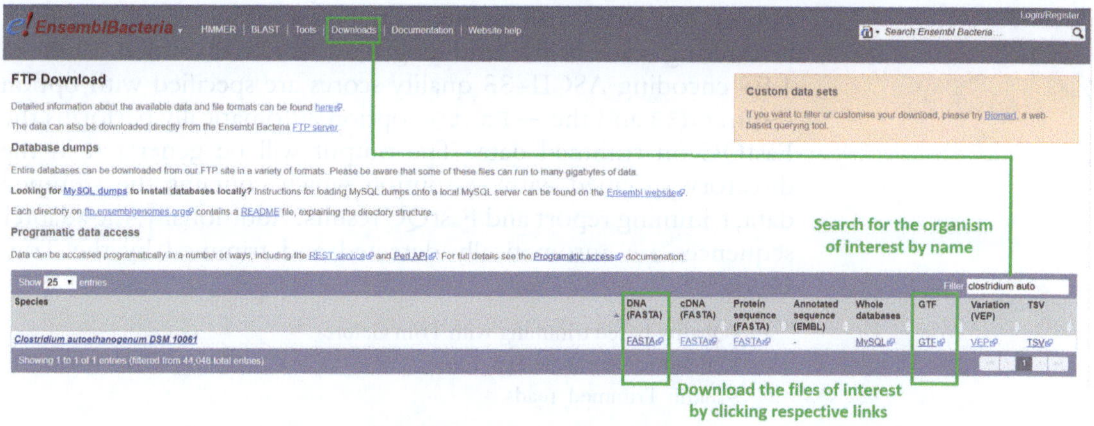

Fig. 1 Schematic illustration for downloading a reference genome

base name. In the example below, we have chosen the base name "`autoethanogenum`"; however, the user can select their choice of base name.

```
#Index the reference genome
$ cd /home/username/exp_run
$ cd reference_genome

$ bowtie2-build genome_ref.fasta autoethanogenum
```

3 Methods

3.1 Running FastQC to Determine the Data Quality Statistics

The following commands run the FastQC tool on specified files and output is generated in the directory specified with the `-o` option. The output directory must exist before running the commands. The successful run generates the FastQC quality statistics report in HTML format and individual result files are available as zip output file.

```
#Determine Quality Statistics for raw data using FastQC
$ cd /home/username/exp_run
$ mkdir FastQC

$ FastQC SRR5069221_1.fastq -o FastQC
$ FastQC SRR5069221_2.fastq -o FastQC
$ FastQC SRR5069222_1.fastq -o FastQC
$ FastQC SRR5069222_2.fastq -o FastQC
```

3.2 Running Trim Galore for Data Quality Trimming

After examining the FastQC quality statistics, we can perform the quality-based trimming of the raw FASTQ files. In this example, we choose the base quality score threshold of 30 specified with `-q` option and minimum length threshold of 50 bp with `--length` option. Any bases with quality score below 30 will be trimmed and if resulting reads below read length 50 will be discarded. Illumina 1.9+ encoding ASCII+33 quality scores are specified with option `--phred33` and the `--FastQC` option automatically performs the FastQC on trimmed data. The output will be generated in the directory specified with `-o` option which contains the trimmed data, trimming report and FastQC results. Additionally, the adapter sequences are automatically detected and trimmed by the Trim Galore.

```
#Quality based trimming with Trim Galore
$ cd /home/username/exp_run
$ mkdir Trimmed_reads

trim_galore -q 30 --length 50 --FastQC --phred33 --paired
./raw_data/SRR5069221_1.fastq ./raw_data/SRR5069221_2.fastq -o
Trimmed_reads
trim_galore -q 30 --length 50 --FastQC --phred33 --paired
./raw_data/SRR5069222_1.fastq ./raw_data/SRR5069222_2.fastq -o
Trimmed_reads
trim_galore -q 30 --length 50 --FastQC --phred33 --paired
./raw_data/SRR5069223_1.fastq ./raw_data/SRR5069223_2.fastq -o
Trimmed_reads
trim_galore -q 30 --length 50 --FastQC --phred33 --paired
./raw_data/SRR5069224_1.fastq ./raw_data/SRR5069224_2.fastq -o
Trimmed_reads
```

An example of per base sequence quality results as generated by FastQC for the file `SRR5069222_2.fastq` before and after quality trimming is shown (Fig. 2).

For each position, a BoxWhisker-type plot is drawn. The elements of the plot are described below:

1. The central red line is the median value.
2. The yellow box represents the inter-quartile range (25–75%).

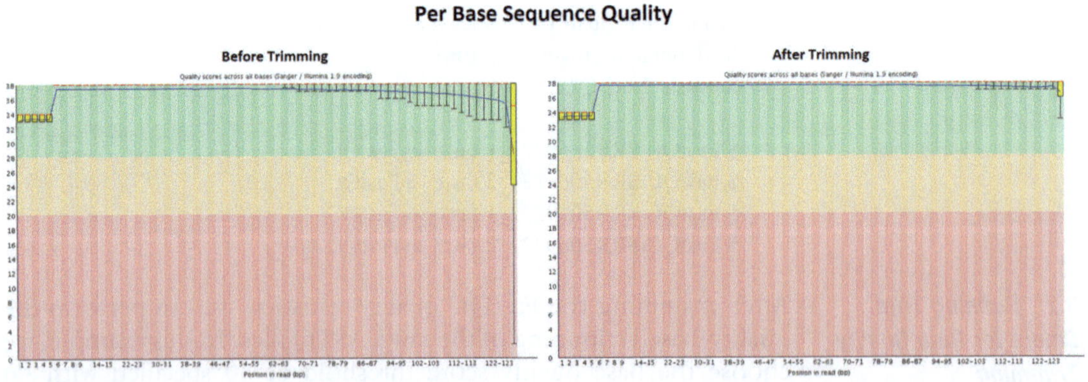

Fig. 2 FastQC quality metric results pre- and post-quality trimming

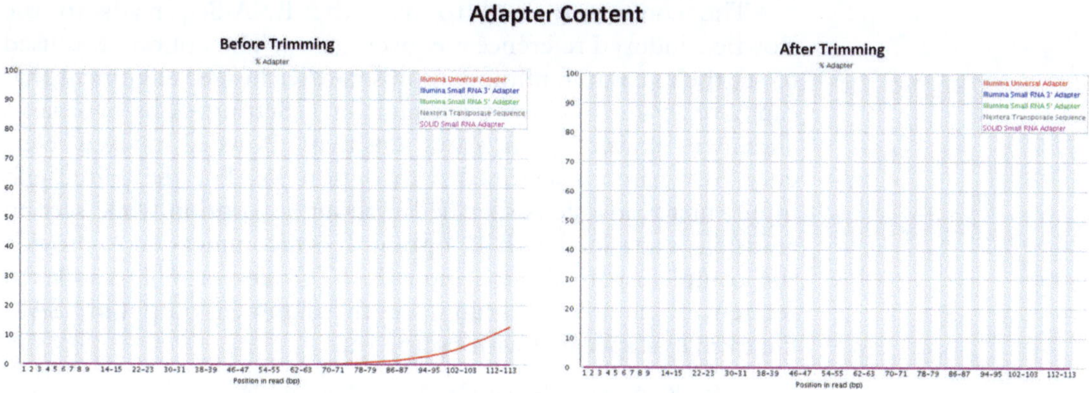

Fig. 3 FastQC adapter assessment

3. The upper and lower whiskers represent the 10 and 90% points.
4. The blue line represents the mean quality.

The *y*-axis on the graph shows the quality scores. The higher the score, the better is the base call quality. The background colors of the graph divide the *y*-axis into very good quality calls (green), calls of reasonable quality (orange), and calls of poor quality (red). The quality of calls on most platforms will degrade as the run progresses, so it is common to see base calls falling into the orange area toward the end of a read. However, after quality trimming, most of the low-quality bases are removed, as shown in the after trimming plot.

An example of adapter contents results for file `SRR5069222_2.fastq` before and after quality trimming as determined by FastQC tool (Fig. 3).

The Kmer Content module of FastQC finds a number of different sources of bias in the library that can include one obvious class of sequences—adapter sequences. It is useful to know if your library contains a significant amount of adapter to assess the need for adapter trimming. The plot itself shows a cumulative percentage count of the proportion of your library that has seen each of the adapter sequences at each position. Once a sequence has been seen in a read, it is counted as being present right through to the end of the read so the percentages you see will only increase as the read length goes on. In above example, Illumina universal adapters present in the original file are automatically detected by Trim Galore and removed after quality trimming.

3.3 Mapping to Reference

After verifying the FastQC results of the quality-trimmed reads, the next step involves mapping the reads to the reference genome. Mapping is the most time-consuming as well as the most memory-intensive process of RNA-Seq workflow.

The commands used to map the RNA-Seq reads to the Bowtie2-indexed reference are given below. The option `-N` is used to set the maximum number of mismatches in seed alignment that can have a value of either 0 (default) or 1. Setting `-N` higher makes alignment slower but increases sensitivity. Bowtie2 indexed reference base name is specified with option `-x`. For the paired-end input, first reads file is specified without any option while second reads file is specified with `-2` option. It is recommended to provide the complete path for the input FASTQ files as well as indexed reference. The successful run stores the alignment output in a SAM file format as specified with option `-S`.

```
#Map RNA-seq reads to the bowtie2-indexed reference
$ cd /home/username/exp_run
$ mkdir Mapping
$ cd Mapping

$ bowtie2 -x
/home/username/exp_run/reference_genome/autoethanogenum -N 1
/home/username/exp_run/Trimmed_reads/SRR5069221_1_val_1.fq -2
/home/username/exp_run/Trimmed_reads/SRR5069221_2_val_2.fq -S
SRR5069221.sam

$ bowtie2 -x
/home/username/exp_run/reference_genome/autoethanogenum -N 1
/home/username/exp_run/Trimmed_reads/SRR5069222_1_val_1.fq -2
/home/username/exp_run/Trimmed_reads/SRR5069222_2_val_2.fq -S
SRR5069222.sam

$ bowtie2  -x
/home/username/exp_run/reference_genome/autoethanogenum -N 1
/home/username/exp_run/Trimmed_reads/SRR506923_1_val_1.fq -2
/home/username/exp_run/Trimmed_reads/SRR5069223_2_val_2.fq -S
SRR5069223.sam

$ bowtie2  -x
/home/username/exp_run/reference_genome/autoethanogenum -N 1
/home/username/exp_run/Trimmed_reads/SRR5069224_1_val_1.fq -2
/home/username/exp_run/Trimmed_reads/SRR5069224_2_val_2.fq -S
SRR5069224.sam
```

In the next few steps, we demonstrate the commands for the conversion of SAM to BAM format, sorting of the BAM file, and generating the mapping statistics. Conversion to BAM format and sorting is often necessary as sorted BAM is a preferred input format for many downstream applications. The `samtools flagstat` command is useful to generate the mapping statistics such as total input reads, total mapped reads, mapped proper pairs and singletons etc.

```
#Process SAM files and generate mapping statistics
$ cd /home/username/exp_run
$ cd Mapping

#Process sample SRR5069221
$ samtools view -bS SRR5069221.sam > SRR5069221.bam
$ samtools sort -n SRR5069221.bam -o SRR5069221_sorted.bam
$ samtools flagstat SRR5069221_sorted.bam >
SRR5069221_flagstat.txt

#Process sample SRR5069222
$ samtools view -bS SRR5069222.sam > SRR5069222.bam
$ samtools sort -n SRR5069222.bam -o SRR5069222_sorted.bam
$ samtools flagstat SRR5069222_sorted.bam >
SRR5069222_flagstat.txt

#Process sample SRR5069223
$ samtools view -bS SRR5069223.sam > SRR5069223.bam
$ samtools sort -n SRR5069223.bam -o SRR5069223_sorted.bam
$ samtools flagstat SRR5069223_sorted.bam >
SRR5069223_flagstat.txt

#Process sample SRR5069224
$ samtools view -bS SRR5069224.sam > SRR5069224.bam
$ samtools sort -n SRR5069224.bam -o SRR5069224_sorted.bam
$ samtools flagstat SRR5069224_sorted.bam >
SRR5069224_flagstat.txt
```

Post mapping statistics for all the sample are provided in the table below (Table 2). Please note that number of reads may differ slightly based on software versions used.

3.4 Generating Read Counts Matrix

In this analysis, we are specifically interested in assignment aligned RNA-Seq reads to corresponding gene features. In the next steps, the number of reads mapped to gene features (represented in GTF file) are counted using HTSeq. Following are the commands used to count the number of RNA-Seq reads aligned to gene features. HTSeq takes the SAM/BMA mapping results and GTF file as input, followed by the extraction of the feature type specified with -t option from the GTF file and counts the number of overlapping reads against each feature using mode specified with -m option. The union is the recommended option for most use cases. The -s option is used to specify whether the data are from a strand-specific assay. The output counts are written to a file specified with > option.

After a successful run, count files for all samples can be combined using a custom script or using Microsoft Excel to generate a

Table 2
Mapping statistics

Sample name	Input reads	Quality trimmed reads	Percentage of reads retained after trimming (%)	Mapped reads	Percentage of mapped reads (%)
SRR5069221	6,796,912	6,684,880	98.35	6,669,326	99.77
SRR5069222	6,707,414	6,604,926	98.47	6,588,878	99.76
SRR5069223	10,773,780	10,537,026	97.80	10,515,794	99.80
SRR5069224	6,694,960	6,562,850	98.03	6,537,130	99.61
SRR5069225	5,621,014	5,497,016	97.79	5,481,341	99.71
SRR5069226	6,298,386	6,131,514	97.35	6,111,706	99.68
SRR5069227	5,792,488	5,707,798	98.54	5,672,751	99.39
SRR5069228	9,531,636	9,382,640	98.44	9,351,996	99.67
SRR5069229	9,585,474	9,393,140	97.99	9,370,047	99.75
SRR5069230	6,588,802	6,416,186	97.38	6,402,302	99.78
SRR5069231	6,225,276	6,116,548	98.25	6,084,640	99.48
SRR5069232	6,801,076	6,710,236	98.66	6,694,672	99.77

master counts matrix. Last five lines in each counts file are summary lines and should be removed from the master counts matrix.

```
#Generate read counts for each gene-feature using HTSeq
$ cd /home/username/exp_run
$ cd Mapping

$ htseq-count -m union -s no -t gene -i gene_id SRR5069221.sam
/home/username/exp_run/reference_genome/genome_ref.gtf >
SRR5069221.counts

$ htseq-count -m union -s no -t gene -i gene_id SRR5069222.sam
/home/username/exp_run/reference_genome/genome_ref.gtf >
SRR5069222.counts

$ htseq-count -m union -s no -t gene -i gene_id SRR5069223.sam
/home/username/exp_run/reference_genome/genome_ref.gtf >
SRR5069223.counts

$ htseq-count -m union -s no -t gene -i gene_id SRR5069224.sam
/home/username/exp_run/reference_genome/genome_ref.gtf >
SRR5069224.counts
```

A snapshot of combined matrix containing read counts for each gene feature across all samples is shown in Table 3. Please note the

Table 3
Example output for matrix containing read counts

ID	LB1	LB2	LB3	LB4	MLB1	MLB2	MB1	MB2	MB3	HB1	HB2	HB3
CAETHG_0001	13	292	271	155	241	183	269	325	336	186	150	173
CAETHG_0002	0	6	0	0	5	5	5	1	6	6	4	6
CAETHG_0003	0	2	9	2	3	5	7	12	5	2	0	10
CAETHG_0004	22	374	597	395	386	350	346	516	551	375	403	448
CAETHG_0005	10	181	399	315	202	207	173	355	388	223	282	314

abbreviations (*LB* Low-Biomass, *MLB* Medium-Low-Biomass, *MB* Medium-Biomass, and *HB* High-Biomass) with replicate numbers (1,2,3,4) used for the samples SRR5069221 to SRR5069232, respectively.

3.5 Differential Gene Expression

The next step is to process the combined counts matrix in R to calculate the differentially expressed genes. Start an R session in the working directory and load the required packages. Step-by-step commands are provided below.

3.5.1 Load the Required R Packages

```
#Load the required R libraries
> library(DESeq2)
> library(ggplot2)
> library(gplots)
> library(pheatmap)
> library(RColorBrewer)
```

3.5.2 Prepare the Counts Data

In following commands, the counts data are read into a variable and converted to an appropriate format. Any genes that have 0 counts across all samples are discarded. When genes have 0 counts in one sample but not in others, the counts were converted from 0 to 1 to avoid having infinite values being calculated for fold change. A final matrix containing nonzero gene counts was used for the further processing.

It is recommended to check the dimensions of the original and nonzero matrix to get an idea of how many genes are discarded. In the current data, original counts matrix contains 4131 genes, while nonzero matrix contains 4114 genes. Therefore, 17 genes were discarded for having 0 counts across all samples.

```
> #Read the counts data
> raw.data <- read.table("counts.TXT",sep="\t", >
header=TRUE,stringsAsFactors=FALSE)

> #read columns 2 to 13 containing counts and assign gene IDs as row
names
> #Note: The number of columns specified are dependent on samples
included in > the counts file
> count_matrix <- raw.data[2:13]
> rownames(count_matrix) <- raw.data$ID

> # Remove rows with zero counts
> count_matrix$rowsum <- rowSums(count_matrix)
> counts_nozero <- subset(count_matrix, count_matrix$rowsum >0)
#subset the data by excluding the column named `rowsum`
> countData <- subset(counts_nozero, select = -c(rowsum))

> #Check the dimensions of the original and non-zero counts matrix
> dim(count_matrix)
> dim(counts_nozero)
```

3.5.3 DESeq2 Preparation

The DESeq2 method uses the negative binomial distribution-based data model and performs specific estimate variance-mean tests. A custom design matrix for the current experiment is created and design is verified by printing the `samples` variable.

```
> #create the custom experiment design matrix
> samples <- data.frame(row.names = colnames(countData))
> samples$condition <- factor(
c(rep("Low",4),rep("Medium_Low",2),rep("Medium",3),rep("HIGH
",3)))

> #Print the samples matrix to verify the design
> samples

##      condition
## LB1       Low
## LB2       Low
## LB3       Low
## LB4       Low
## MLB1 Medium_Low
## MLB2 Medium_Low
## MB1     Medium
## MB2     Medium
## MB3     Medium
## HB1       HIGH
## HB2       HIGH
## HB3       HIGH
```

The function DESeqDataSetFromMatrix is used when a counts matrix is available from the external tool such as HTSeq. The following command shows the preparation of DESeq2 object using counts data and custom experiment design. The `design = ~ conditions` refer to the condition column in the `samples` design matrix and denotes the sample groups.

```
> # Prepare DESeq2 Object
> dds<- DESeqDataSetFromMatrix(countData = countData,
               colData = samples,
               design = ~ condition)
```

The standard differential expression analysis steps are wrapped into a single function `DESeq`. In DESeq2, differential expression is determined by application of Empirical Bayes shrinkage for dispersion estimation and Wald test is used for the significance testing. The test determines the differentially expressed genes with *P-value* and adjusted P-values of false discovery rate (FDR) to correct for multiple tests.

```
> # Calculate results
> dds <- DESeq(dds)
```

3.5.4 Unsupervised Data Exploration

Data quality assessment and quality control (i.e., the removal of inferior quality data) are essential steps of any data analysis. These steps are typically performed prior to the differential expression testing. Exploratory plots provide information about general attributes of the data such as how samples cluster and whether any libraries are suspect and should be removed prior to performing differential expression analysis. Please note that previous FastQC steps were used for the quality trimming of individual sample data, while exploratory plots allow evaluating all sample data together and determine outliers.

```
> #rld is preferable for the varying size factors and in most cases size

> factors vary a lot
> rld <- rlog(dds, blind= FALSE)
```

Principal Component Analysis (PCA) plot shows the samples in the two-dimensional plane spanned by their first two principal

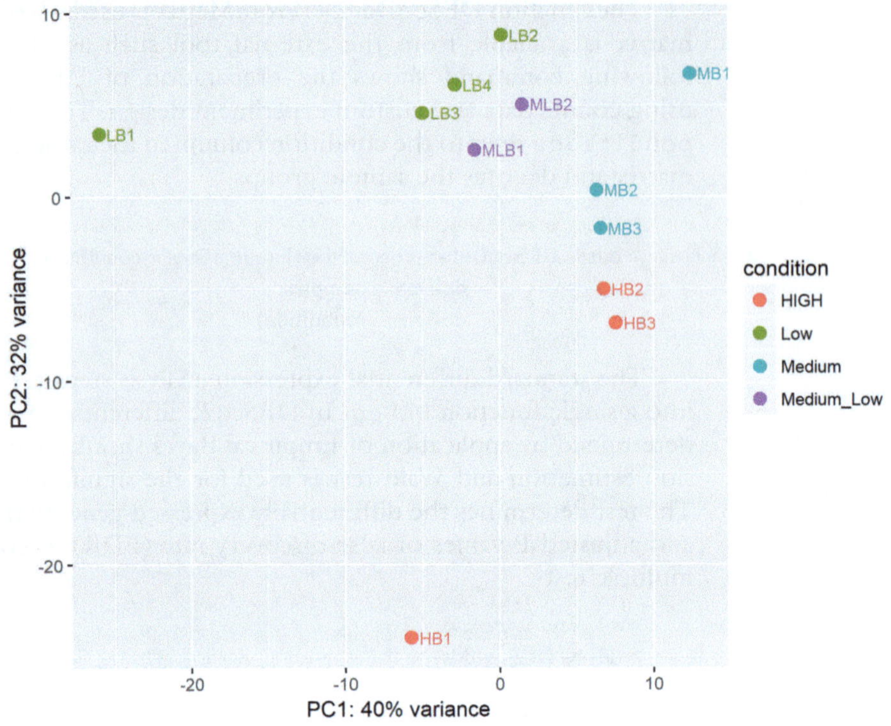

Fig. 4 Principal component analysis

components (Fig. 4). This type of plot is useful for visualizing the overall effect of experimental covariates and batch effects.

```
> #create a PCA plot
> pca <- plotPCA(rld, intgroup=c("condition"), returnData=TRUE)
> percentVar1 <- round(100 * attr(pca, "percentVar"))

> p1 <- ggplot(pca, aes(PC1, PC2,color= condition,
theme(legend.position = "none"),label = rownames(pca))) +
>   geom_point(size=0.5) +
>   xlab(paste0("PC1: ",percentVar1[1],"% variance")) +
>   ylab(paste0("PC2: ",percentVar1[2],"% variance"))

> p1 + geom_point(size = 3) + geom_text(hjust =0, nudge_x = 0.5,
size=3) + theme(legend.position = "right")
```

A heatmap of sample-to-sample Euclidian distances can be prepared with the DESeq2 transformed data after regularized log transformations (Fig. 5). This is useful to determine the sample-to-sample clustering and possible outliers.

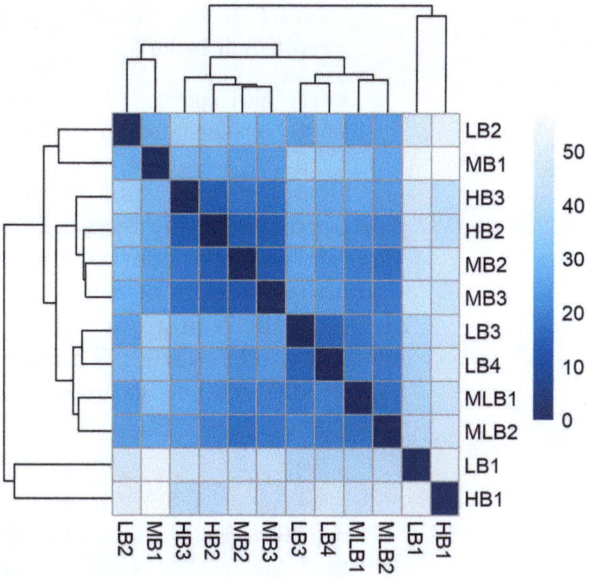

Fig. 5 Cluster analysis

```
> #generate a heatmap of sample-to-sample distances

> #prepare distance matrix
> sampleDists <- dist(t(assay(rld)))
> sampleDistMatrix <- as.matrix(sampleDists)

> #select the color pattern for heatmpap
> colors <- colorRampPalette( rev(brewer.pal(9, "Blues")) )(255)

> #generate heatmap
> pheatmap(sampleDistMatrix,
      clustering_distance_rows=sampleDists,
      clustering_distance_cols=sampleDists,
      col=colors,
      show_rownames = TRUE,
      show_colnames = TRUE)
```

3.5.5 Differential Gene Expression with DESeq2

When multiple groups or conditions (e.g., Low-Biomass, Medium-Low-Biomass, Medium-Biomass, and High-Biomass) are defined in the counts matrix, the `contrast` argument is used to test the differences between the groups. The names for contrasting should be exact as defined in the DESeq2 object. The `resultsNames` function can be used to determine the exact names within DESeq2 object.

```
> #Check the resultNames in DESeq2 object
> resultsNames(dds)

## [1] "Intercept"         "conditionHIGH"      "conditionLow"
## [4] "conditionMedium"    "conditionMedium_Low"
```

Next, the `results` function of the DESeq2 package performs independent filtering by using the mean of normalized counts as a filter statistic. The contrast argument of results function is used to extract test results of log2 fold changes of interest. Ensure the names in the contrast argument are exactly the same as provided by the `resultsNames`. The first few rows of the result table containing differentially expressed genes can be viewed with `head()` command.

```
> #Calculate differentially expressed genes in comparison Medium-
Low-Biomass vs Low-Biomass
> MLB_vs_LB <- results(dds, contrast =
list("conditionMedium_Low","conditionLow"))

> #check first few rows of the output
> head(MLB_vs_LB)

## log2 fold change (MAP): conditionMedium_Low vs conditionLow
## Wald test p-value: conditionMedium_Low vs conditionLow
## DataFrame with 6 rows and 6 columns
##            baseMean log2FoldChange     lfcSE      stat   pvalue
##           <numeric>      <numeric> <numeric> <numeric>
<numeric>
## CAETHG_0001 182.407786    0.15329614 0.3212131  0.4772413
0.6331903
## CAETHG_0002   3.188140    0.38466245 0.3561228  1.0801400
0.2800799
## CAETHG_0003   3.518217    0.12805483 0.3747426  0.3417141
0.7325660
## CAETHG_0004 331.247969    0.09401265 0.2476970  0.3795470
0.7042817
## CAETHG_0005 205.169750   -0.06761588 0.2896284 -0.2334573
0.8154063
## CAETHG_0006  53.288699   -0.13523477 0.3289662 -0.4110902
0.6810064
##               padj
##          <numeric>
## CAETHG_0001 0.9983943
## CAETHG_0002        NA
## CAETHG_0003        NA
## CAETHG_0004 0.9983943
## CAETHG_0005 0.9983943
## CAETHG_0006 0.9983943
```

A simple `subset` commands in R can be used to filter the result table to determine significant genes with padj (i.e., False Discovery Rate) scores less than 0.05, up-regulated genes with log2Fold-Change greater than 1 and down-regulated genes with log2Fold-Change less than −1. Please note that the user can select different cutoff of their choice to determine the significant, up- and down-regulated genes.

```
> #Determine the significant differentially expressed genes (padj <=
0.05)
> MLB_vs_LB$Gene_ID = rownames(MLB_vs_LB)
> MLB_vs_LB_significant <- subset(MLB_vs_LB,
MLB_vs_LB$padj <= 0.05)

> #Determine the up-regulated (log2foldchange > 1) and down -
regulated (log2FC < -1) genes
> MLB_vs_LB_up  <- subset(MLB_vs_LB_significant,
MLB_vs_LB_significant$log2FoldChange > 1)
> MLB_vs_LB_down <- subset(MLB_vs_LB_significant,
MLB_vs_LB_significant$log2FoldChange < -1)
```

The log2FoldChanges of the genes within specific comparison can be visualized with MA plots. The MA plot shows the distribution of the mean of normalized counts (*x*-axis) and log2fold change (*y*-axis) of the genes (Fig. 6).

```
> #Create MA plot for the comparison Medium-Low-Biomass vs Low-
Biomass
> DESeq2::plotMA(MLB_vs_LB, main="MA Plot", ylim=c(-3,3),
cex=.8)
> abline(h=c(-1,1), col="dodgerblue", lwd=2)
```

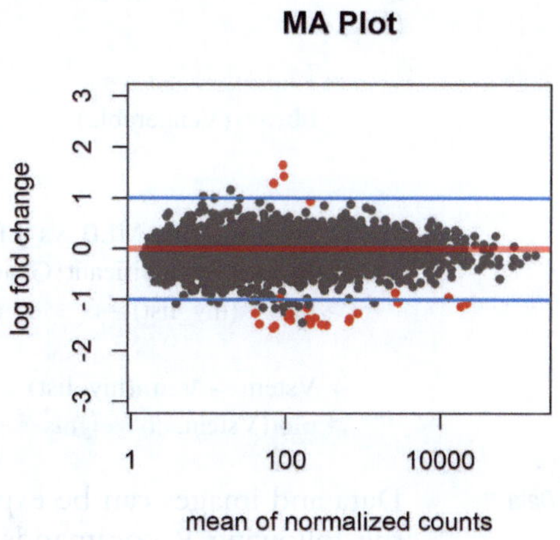

Fig. 6 MA plot for distribution analysis. Genes with high fold-change values (log2FC > 1 and log2FC < −1) are shown beyond blue lines. Significant genes (*p*-adj < 0.05) are shown in red

Similarly, to perform a comparison across different groups, a contrast argument can be altered as shown below:

```
> #Calculate differentially expressed genes in comparison Medium-Biomass vs Low-Biomass
> MB_vs_LB <- results(dds, contrast = list("conditionMedium","conditionLow"))
> MB_vs_LB$Gene_ID = rownames(MB_vs_LB)
> MB_vs_LB_significant <- subset(MB_vs_LB, MB_vs_LB$padj <= 0.05)
> MB_vs_LB_up  <- subset(MB_vs_LB_significant, MB_vs_LB_significant$log2FoldChange > 1)
> MB_vs_LB_down <- subset(MB_vs_LB_significant, MB_vs_LB_significant$log2FoldChange < -1)

> #Calculate differentially expressed genes in comparison High-Biomass vs Low-Biomass
> HB_vs_LB <- results(dds, contrast = list("conditionHIGH","conditionLow"))
> HB_vs_LB$Gene_ID = rownames(HB_vs_LB)
> HB_vs_LB_significant <- subset(HB_vs_LB, HB_vs_LB$padj <= 0.05)
> HB_vs_LB_up  <- subset(HB_vs_LB_significant, HB_vs_LB_significant$log2FoldChange > 1)
> HB_vs_LB_down <- subset(HB_vs_LB_significant, HB_vs_LB_significant$log2FoldChange < -1)
```

3.5.6 Creating a Venn Diagram

It may be useful to visualize the number of overlapping genes across three comparisons. This could be easily achieved using the `Vennerable` library in R. The following commands show how to generate a Venn diagram of significant genes across comparisons (Fig. 7).

```
> #load the required library
> library(Vennerable)

> #generate a list of significant genes
> my_list <- list(MLB_vs_LB_significant$Gene_ID, MB_vs_LB_significant$Gene_ID, HB_vs_LB_significant$Gene_ID)
> names(my_list) = c("Medium-Low", "Medium", "High")

> Vstem <- Venn(my_list)
> plot(Vstem, doWeights = FALSE)
```

3.5.7 Exporting Data Through R

Data and images can be exported in the appropriate format using the following R commands. After the successful run, the output should be generated in the files `MA_plot.PNG` and `MLB_vs_LB.txt`, respectively.

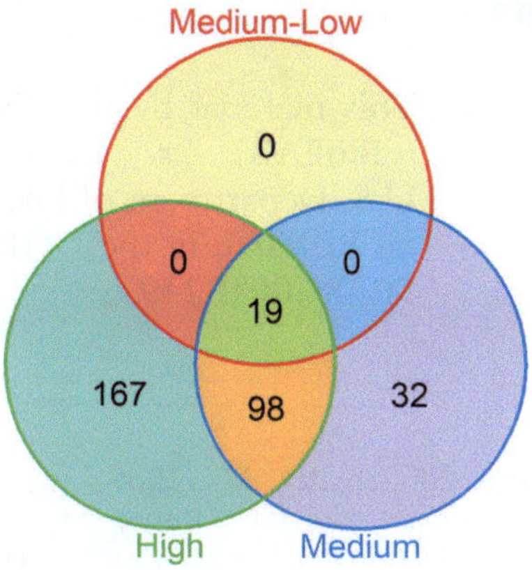

Fig. 7 Venn analysis of differential expression results

```
> # store the MA plot as PNG file
> png(filename ="MA_plot.PNG", width = 600, height = 600,
pointsize = 12,  bg = "white", res =100)
> DESeq2::plotMA(MLB_vs_LB, main="MA Plot", ylim=c(-3,3),
cex=.8)
> abline(h=c(-1,1), col="dodgerblue", lwd=2)
> dev.off()

> # Export the significant differentially expressed genes from
comparison Medium-Low-Biomass vs Low-Biomass
> write.table(MLB_vs_LB_significant, file="MLB_vs_LB.txt",
row.names = FALSE, quote=FALSE, sep='\t')
```

4 Notes

As a general guideline, it is suggested that a minimum of five million reads is used for bacterial RNA-seq experiments. We note that upon outlier detection, samples can be excluded, and statistical tests can be rerun. Depending upon the experimental system, results, and biological questions, thresholds for fold-change and statistical significance can be altered to derive appropriate analyses for further interpretation. We present one pipeline, and other tools could be substituted into different parts of the pipeline. Readers should see the original publication for interpretation of physiological differences between conditions [10]. In this example we used the publicly available wild-type reference genome previously described [17].

Acknowledgments

This material by the *Clostridium* foundry for biosystems design (cBioFAB) is based upon work supported by the U.S. Department of Energy, Office of Biological and Environmental Research in the DOE Office of Science under Award Number DE-SC0018249.

References

1. Wang Z, Gerstein M, Snyder M (2009) RNA-Seq: a revolutionary tool for transcriptomics. Nat Rev Genet 10:57
2. Hör J, Gorski SA, Vogel J (2018) Bacterial RNA biology on a genome scale. Mol Cell 70:785–799
3. Anders S, McCarthy DJ, Chen Y et al (2013) Count-based differential expression analysis of RNA sequencing data using R and Bioconductor. Nat Protoc 8:1765
4. Ozsolak F, Milos PM (2010) RNA sequencing: advances, challenges and opportunities. Nat Rev Genet 12:87
5. Manga P, Klingeman DM, Lu T-YS et al (2016) Replicates, read numbers, and other important experimental design considerations for microbial RNA-seq identified using *Bacillus thuringiensis* datasets. Front Microbiol 7:794
6. Dillies M-A, on behalf of The French StatOmique Consortium, Rau A et al (2013) A comprehensive evaluation of normalization methods for Illumina high-throughput RNA sequencing data analysis. Brief Bioinform 14:671–683
7. Gierliński M, Cole C, Schofield P et al (2015) Statistical models for RNA-seq data derived from a two-condition 48-replicate experiment. Bioinformatics 31:3625–3630
8. Mi G, Di Y, Schafer DW (2015) Goodness-of-fit tests and model diagnostics for negative binomial regression of RNA sequencing data. PLoS One 10:1–16
9. Miller CA, Hampton O, Coarfa C et al (2011) ReadDepth: a parallel R package for detecting copy number alterations from short sequencing reads. PLoS One 6:1–7
10. Valgepea K, de Souza Pinto Lemgruber R, Meaghan K et al (2017) Maintenance of ATP homeostasis triggers metabolic shifts in gas-fermenting Acetogens. Cell Syst 4:505–515.e5
11. Liew F, Martin ME, Tappel RC et al (2016) Gas fermentation—a flexible platform for commercial scale production of low-carbon-fuels and chemicals from waste and renewable Feedstocks. Front Microbiol 7:694
12. Heijstra BD, Leang C, Juminaga A (2017) Gas fermentation: cellular engineering possibilities and scale up. Microb Cell Factories 16:60
13. FastQC, https://www.bioinformatics.babraham.ac.uk/projects/fastqc/
14. Langmead B, Salzberg SL (2012) Fast gapped-read alignment with bowtie 2. Nat Methods 9:357–359
15. Anders S, Pyl PT, Huber W (2015) HTSeq—a Python framework to work with high-throughput sequencing data. Bioinformatics 31:166–169
16. Love MI, Huber W, Anders S (2014) Moderated estimation of fold change and dispersion for RNA-seq data with DESeq2. Genome Biol 15:550
17. Brown SD, Nagaraju S, Utturkar S et al (2014) Comparison of single-molecule sequencing and hybrid approaches for finishing the genome of *Clostridium autoethanogenum* and analysis of CRISPR systems in industrial relevant clostridia. Biotechnol Biofuels 7:1–18

Chapter 9

Measuring Biomass-Derived Products in Biological Conversion and Metabolic Process

Chang Geun Yoo, Yunqiao Pu, and Arthur J. Ragauskas

Abstract

Biomass can be converted to various types of products in biological and metabolic processes. For an in-depth understanding of biomass conversion, quantitative and qualitative information of products in these conversion processes are essential. Here we introduce analytical techniques including high-performance liquid chromatography (HPLC), gas chromatography (GC), gas chromatography-mass spectrometry (GC-MS), and nuclear magnetic resonance (NMR) for biomass-based products characterization in biological and metabolic processes.

Key words Biomass-based products, Metabolic process, High-performance liquid chromatography, Gas chromatography, Nuclear magnetic resonance

1 Introduction

Biomass is a great feedstock in many biological processes. Various types of products are produced during biological and metabolic processes because of its heterogeneous and complicated structure and variety depending on the feedstock and environments. For these reasons, selecting and applying suitable quantitative and qualitative analysis methods are important to understand and evaluate the products and by-products in the biological and metabolic process.

Carbohydrates are major components of biomass and exist as different forms such as cellulose, hemicellulose, starch, and sucrose. These carbohydrates can be converted to fermentable sugars for acetone, butanol, ethanol, and other fermentation products. In addition, these carbohydrates can be directly used in many food and pharmaceutical applications. Qualification and quantification of carbohydrates in the products are essential in biological and metabolic processes. Colorimetric techniques such as anthrone, phenol-sulfuric acid, and 3,5-dinitrosalicylic acid (DNS) methods have been traditionally applied for carbohydrate

Michael E. Himmel and Yannick J. Bomble (eds.), *Metabolic Pathway Engineering*, Methods in Molecular Biology, vol. 2096, https://doi.org/10.1007/978-1-0716-0195-2_9, © Springer Science+Business Media, LLC, part of Springer Nature 2020

determination [1–4]. High-performance liquid chromatography (HPLC) was introduced to overcome the technical barriers of the colorimetric methods [5, 6]. Acetone, butanol, and ethanol are well-known fermentation products from carbohydrates. These products can be produced together in the acetone-butanol-ethanol (ABE) fermentation process and are also available individually. Ethanol is the most popular biofuel from diverse biomass, including cellulosic and starch-based feedstock. It can be blended with gasoline and applied as transportation fuel without engine modification [7]. Butanol is another fermentation product from biomass by some *Clostridium* species [8, 9]. Compared to bioethanol, it has higher energy content, lower volatility, less hydroscopic, and less corrosiveness [10, 11]. Acetone is also produced as a co-product from ABE fermentation. These chemicals are not only used as they are but can also be converted to hydrocarbons through a catalytic upgrading process [12].

Recently, increased interests on biomass utilization have significantly broadened its applications. Besides the aforementioned biomass-based products, conversion of biomass to value-added products such as lactic acid, astaxathin, and lipids by biological conversion processes have been studied. Lactic acid is an important building block for many applications in pharmaceuticals, cosmetics, and food. It can be used as a precursor for polylactic acid (PLA) and is also used in pharmaceutical and cosmetic industries as a starting material [13]. In addition, it can be applied as a food additive [14]. Although it can be produced by both fermentation and chemical synthesis, fermentation is a predominant approach for industrial lactic acid production. Astaxanthin is also a high-value fermentation product [15]. It is used as a dietary supplement for human and as an animal feed additive [16, 17]. With its growing market, it is considered as a value-added coproduct from biomass. Lipids including fatty acids, phospholipids, sterols, sphingolipids, terpenes, and others are a ubiquitous group of organic compounds, which are soluble in organic solvents but insoluble in water [18]. Accumulation of lipids using oleaginous bacteria, engineered algae, and yeast has been reported for biofuel application [19]. Some other chemicals such as acetic acid, formic acid, levulinic acid, furfural, HMF and mono-aromatics are possibly observed from biological product streams either as a coproducts by the conversion processes or as impurities generated from the preprocess like pretreatment. In this chapter, we present the HPLC, GC, GC-MS, and NMR analysis of the products in the biological conversion and the metabolic process.

2 Materials

Prepare sample and standard solutions using deionized (DI) water (18 MΩ-cm at room temperature) and analytical grade chemicals at room temperature (unless indicated otherwise). Store the prepared samples and standards at 4 °C in the refrigerator until operating the analysis. Generated waste materials during the standards and sample preparations should be disposed according to the waste disposal regulations.

3 Methods

3.1 High-Performance Liquid Chromatography (HPLC) Method

3.1.1 HPLC Analysis for Carbohydrates

Calibration Standards Preparation

1. Prepare internal standard solution: Dissolve fucose in DI water.
2. Prepare external standard solution: Dissolve recommended amounts of each carbohydrate (glucose, xylose, galactose, arabinose, and mannose) in DI water.
3. Mix the external solution at each concentration with the internal standard solution.

Sample Preparation

1. Take 1 mL of aliquot in the reactor into 2 mL microcentrifuge tube.
2. Prepare water bath at ~100 °C and locate the tubes for 10 min to deactivate enzymes/microorganisms in the samples (if necessary).
3. Cool down the sample in ice bath.
4. Centrifuge the samples at 12,000 × *g* for 10 min.
5. Take 0.5 mL of supernatant from the centrifuged samples and filter using 0.2 μm nominal pore size nylon syringe filters.
6. Dilute the aliquots with DI water depending on the anticipated carbohydrates concentration.
7. Take the diluted sample solution and mix with the fucose internal standard solution at the same ratio of standards.
8. Prepare substrate blank, which contains the same blank solution, substrate, and antimicrobials without enzymes.
9. Prepare enzyme blank, which contains the same blank solution, enzymes, and antimicrobials without substrate (*see* **Note 1**).

Analysis

1. Prepare calibration curves of each carbohydrate with the standard solutions by operating high-performance liquid chromatography equipped with carbohydrate column and electrochemical detector (or refractive index (RI) detector) (*see* **Note 2**).

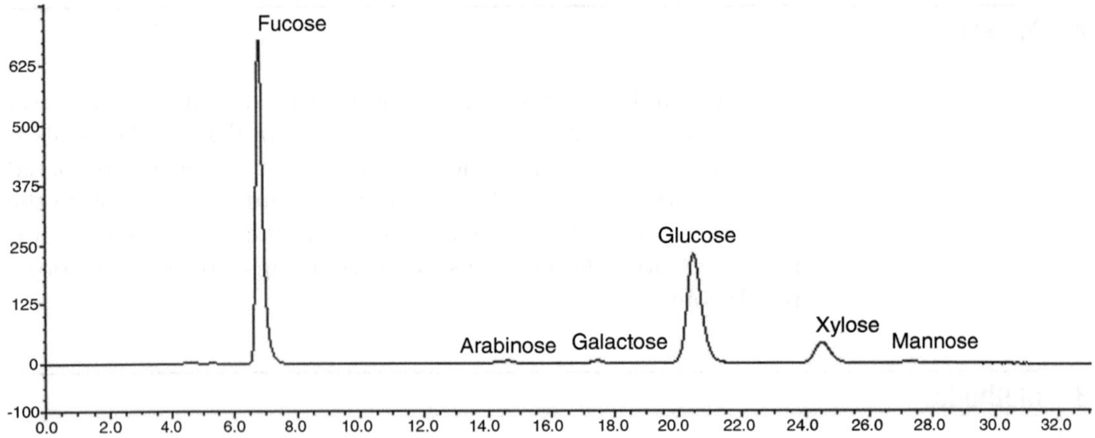

Fig. 1 HPLC peaks detected by electrochemical detector for carbohydrates and internal standard (fucose) in sample solution

2. Run the sample solutions (Fig. 1) and blanks (substrate blank and enzyme blank) under the same HPLC operation conditions.
3. Identify and quantify the carbohydrates in the sample solution by comparing the peaks in the sample solution to those in the standard solutions and blanks.

3.1.2 HPLC Analysis for Fermentation Products

Calibration Standards Preparation

1. Dissolve the expected fermentation products such as acetone, 200 proof ethanol, n-butanol, and lactic acid (analytical grade) in DI water at room temperature.
2. Dilute five different concentrations of external standard solution with DI water.

Sample Preparation

1. Take 1 mL of aliquot in the reactor into 2 mL microcentrifuge tube.
2. Centrifuging the samples at 12,000 × *g* for 15 min.
3. Take 0.5 mL of supernatant from the centrifuged samples and filter using 0.2 μm nominal pore size nylon syringe filters.

Analysis

1. Prepare calibration curves of each product with the prepared standards by operating HPLC equipped with an organic acid analysis column and RI detector (or UV detector) (*see* **Note 3**).
2. HPLC operating conditions: 5 mM H_2SO_4 as mobile phase with 0.6 mL/min flow rate, column temperature is 65 °C [20].
3. Identify and quantify the products by comparing the peaks in the samples to those in the standards (Fig. 2).

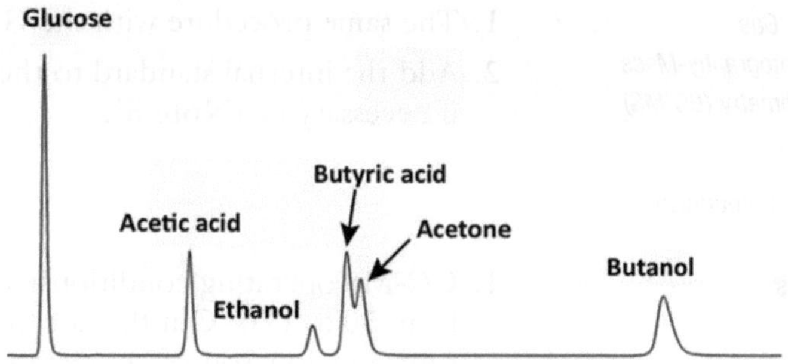

Fig. 2 HPLC peaks detected by RI detector for ABE fermentation products. (Reproduced from [20] with permission from The Royal Society of Chemistry)

3.2 Gas Chromatography (GC) Analysis for Fermentation Products

3.2.1 Gas Chromatography with Flame Ionization Detector (GC-FID) Analysis

Calibration Standards Preparation

1. Dissolve the expected fermentation products (analytical grade) in tributyrin at room temperature.
2. Dilute five different concentrations of standard solution with tributyrin.

Sample Preparation

1. Add extractant to the bioreactor. Take aliquot from the extractant phase of the reactor into micro-centrifuge tube.
2. Centrifuge the tubes at 9400 × *g* for 3 min.
3. Collect and filter the extractant into GC vial using a 0.22 μm syringe filter. Store the filtered extractant at −20 °C prior to conducting GC-FID analysis.

Analysis

1. Prepare calibration curves of each product with the prepared standards by operating GC equipped with a flame ionization detector (FID) and capillary column.
2. GC-FID operating conditions: The oven temperature increases from 35 to 150 °C at the rate of 10 °C/min and ramp to 300 °C at 20 °C/min. Injector temperature and detector temperatures are 260 °C and 280 °C. The split ratio is set at 60. The Factor Four capillary column (VF-5 ms) is used [12] (*see* **Note 4**).
3. Identify and quantify the products by comparing the peaks in the samples to those in the standards.

3.2.2 Gas Chromatography-Mass Spectrometry (GC-MS) Analysis

Sample Preparation

1. The same procedure with the GC-FID analysis can be applied.
2. Add the internal standard to the prepared samples in GC vials, if necessary (*see* **Note 5**).

Analysis

1. GC-MS operating conditions: The oven temperature increases from 50 to 170 °C at the rate of 10 °C/min. Injector temperature and ion source temperatures are 250 °C and 200 °C, respectively. Carrier gas is a helium gas and set at 1.0 mL/min. The split ratio is set at 25. The HP-INNOWax column (19091N-233) is used [21] (*see* **Note 4**).
2. Identify the products with NIST chemical library (Fig. 3).
3. If necessary, compare the peak area of the target compound in the samples to the known concentration internal standards for quantification.

3.3 Extraction and Quantification of Astaxanthin Using UV/Vis Spectrometer

3.3.1 Calibration Standards Preparation

1. Dissolve astaxanthin (analytical grade) in acetone at room temperature.
2. Dilute five different concentrations of external standard solution with acetone.

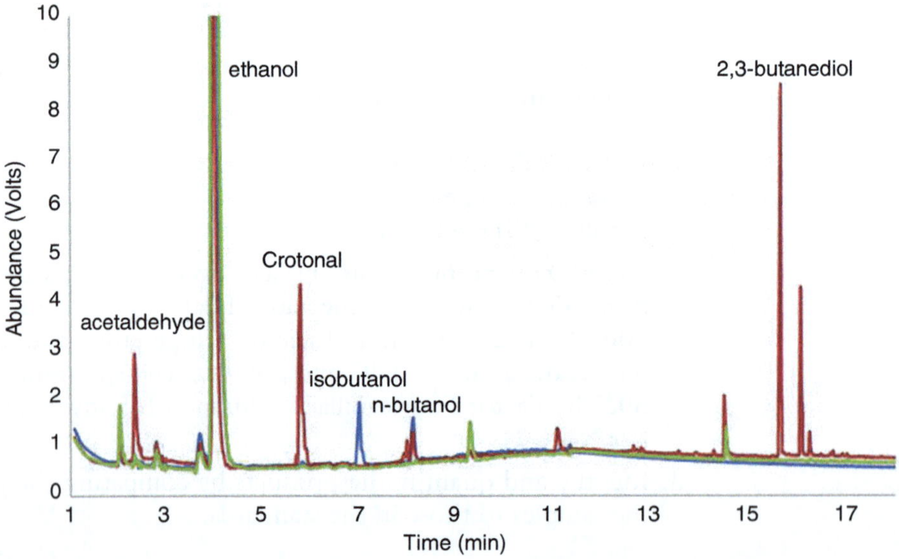

Fig. 3 GC-MS chromatograms of butanol fermentation products. (Reproduced from [22])

3.3.2 Sample Preparation and Analysis

1. Take 1 mL of aliquot in the reactor into glass test tube with 3 mL of DI water.
2. Centrifuge the samples at 12,000 × *g* for 5 min and remove the supernatant.
3. Add glass beads and 1.5 mL of acetone to the test tube.
4. Vortex the test tube for 1 min and sonicate for 5 min.
5. Centrifuge the samples at 12,000 × *g* for 5 min.
6. Transfer the supernatant to a quartz cuvette and read at 480 nm using UV/Vis spectrophotometer [15].
7. Use pure acetone as a blank measurement and obtain a calibration curve from standards.
8. Quantify the products using the calibration curve.

3.4 Extraction and Characterization of Lipids Using GC-MS

3.4.1 Sample Preparation

1. Separate solid residues after fermentation by vacuum filtration.
2. Wash the solid residues with DI water and freeze dry.
3. Dissolve the freeze-dried residues in the mixture of chloroform, methanol, and concentrated sulfuric acid and incubate at 100 °C for 140 min to obtain fatty acid methyl esters (FAME) [19].
4. Add DI water and vortex for 1 min.
5. Recover organic phase after phase separation.
6. Dilute with chloroform before GC analysis.

3.4.2 Analysis

1. GC-MS operating conditions: The oven temperature increases from 50 to 200 °C at the rate of 15 °C/min, to 250 °C at the rate of 10 °C/min, and to 300 °C at the rate of 25 °C/min (Hold at 50 °C for 2 min, 200 °C for 5 min, 250 °C for 2 min, and 300 °C for 4 min). Injector temperature and ion source temperatures are 250 °C and 200 °C. Carrier gas is a helium gas and the HP-5MS column is used [19].
2. Identify the products with the NIST chemical library and external standard (the Supelco® 37 Component FAME Mix, CRM47885) (Fig. 4).

3.5 Nuclear Magnetic Resonance (NMR) Analysis of Fermentation Products

3.5.1 Sample Preparation

1. The fermentation suspensions are taken into micro-centrifuge tube and centrifuged at 13,000 × *g* for 5 min at room temperature resulting in a cell-free supernatant. The supernatants are then collected (*see* **Note 6**).
2. The clear solution (0.54 mL) is transferred into a NMR tube (5.0 mm O.D., Wilmad). Deuterium oxide (D_2O, 60 μL) containing 3-(trimethylsilyl)propionic-2,2,3,3-d_4-acid (TSP, sodium salt, 30 ng/mL) is added, and tubes are mixed with vortexing for 1 min (*see* **Note 7**).

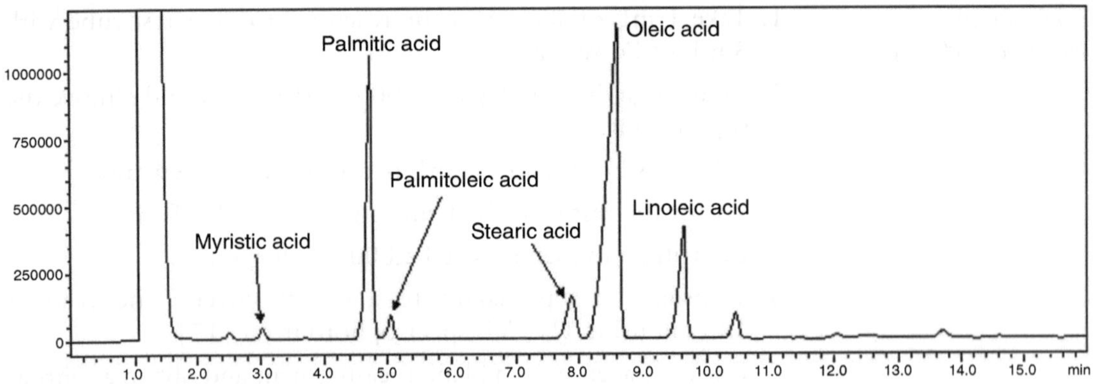

Fig. 4 GC chromatograms of FAME of *Mortierella isabellina*

3.5.2 ^{1}H NMR

1. Proton NMR experiments are run on a Bruker Avance-III 500 MHz HD spectrometer with a cryoprobe at controlled temperature of 300 K using a SampleJet automatic sample changer.
2. The sample is locked to the deuterated solvent. Tuning and matching of ^{1}H nucleus is adjusted. The sample is shimmed using Bruker Topshim.
3. The spectra are acquired with optimized parameter sets. Typical ^{1}H data acquisition parameters are as follows: 90° pulse, 15 ppm spectral width, 32–128 scans. Relaxation delay is set to 4 s. Water suppression pulse is employed for solvent suppression.
4. The spectra are processed using Bruker Topspin software or available commercial software such as MNova (Mestrelab Research S.L., Santiago de Compostela, Spain) and/or Chenomix (Chenomix, Edmonton, Alberta, Canada) software. Manual spectra phasing and baseline corrections are performed.
5. The products and functional groups in the samples are identified by comparing the chemical shifts of peaks and multiplicities with database of known metabolites (for example, publicly available metabolite libraries from the University of Wisconsin and the Magnetic Resonance Metabolomics Database (ESMRMB), Basel, Switzerland) (Fig. 5).
6. Peak integration of metabolite peaks is carried out to quantify the relative products contents relative to TSP.

3.5.3 ^{13}C-NMR

1. ^{13}C-NMR experiments are run on a Bruker Avance-III 500 MHz HD spectrometer equipped with a dual C/H-cryoprobe operated at frequency of 125.18 MHz for the ^{13}C nucleus (*see* **Note 8**).

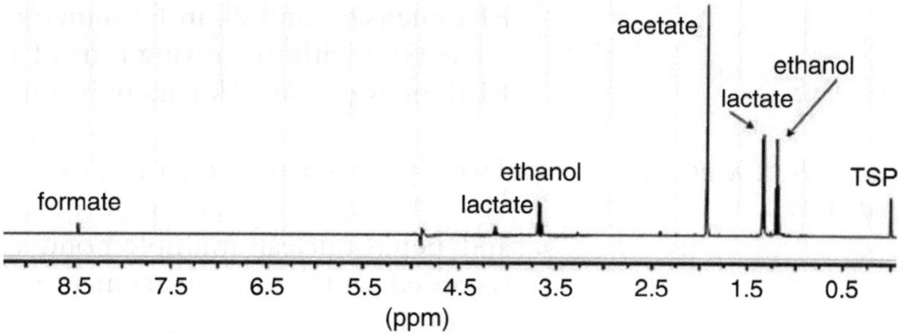

Fig. 5 Proton NMR spectrum of fermentation media from suspensions of *Synechococcus* sp. PCC 7002. (Reprinted from John Wiley & Sons, Inc.) [23]

2. The sample is locked to the deuterated solvent. Tuning and matching of both ^{1}H and ^{13}C nuclei are adjusted, and the sample is shimmed using Bruker Topshim.
3. Both qualitative and quantitative ^{13}C spectra can be carried out for identification and quantification purpose, respectively. The spectra are collected using a 230 ppm spectra width. For qualitative spectrum, the acquisition is performed using a 30° pulse with a gated decoupling pulse sequence (Bruker pulse zgpg) to enhance the signal intensities at 300 K. A short pulse delay of 1-s is used with a minimal of 12,288 scans accumulated for each sample. For quantitative analysis, a 90° pulse with an inversed gated decoupling pulse sequence is used with the pulse delay $\geq 5T_1$ to ensure the quantitative nature of signals (*see* **Note 9**). The number of scans can vary from 5k to 20k depending on the sample and interested metabolites concentration.
4. The spectra are processed using Bruker Topspin software or available commercial software. The recorded free induction decay signals are Fourier transformed with a manual phase correction applied. Chemical shifts calibration and baseline correction are usually needed. Identification of peaks is performed based on the available spectra database of metabolites.

3.5.4 ^{1}H-^{1}H DQF-COSY and TOCSY NMR

1. The double quantum filtered correlation spectroscopy (DQF-COSY) and total correlation spectroscopy (TOCSY) NMR experiments are run on a Bruker Avance-III 500 MHz HD spectrometer with a cryoprobe at controlled temperature of 300 K.
2. The sample is locked to the deuterated solvent. Tuning and matching of ^{1}H nucleus is adjusted, and the sample is shimmed using Bruker Topshim.
3. DQF-COSY and TOCSY experiments are performed using Bruker standard pulse sequences with ^{1}H spectra width of 12 ppm, relaxation delay of 2 s, and 64–128 scans. For DQF-COSY experiment, the data points are set to be 256 in

F1 dimension and 2k in F2 dimension. TOCSY experiment is carried out with the mixing time of 100 ms, 128 data points in F1 dimension, and 1k data points in F2 dimension.

3.5.5 ^{1}H-^{13}C NMR HSQC and HMBC

1. Two-dimensional heteronuclear correlation experiments, ^{1}H–^{13}C heteronuclear single quantum coherence (HSQC), and heteronuclear multiple-bond connectivity (HMBC) are recorded on the same instrument with controlled temperature of 300 K.
2. The sample is locked to the deuterated solvent. Tuning and matching of ^{1}H nucleus is adjusted, and the sample is shimmed using Bruker Topshim.
3. The experiments are performed using Bruker standard pulse sequences with ^{1}H spectral width of 12 ppm and ^{13}C spectral widths of 230 ppm. The condition for HSQC is as following: 256 complex points in the indirect dimension and 1024 data points in F2 dimension, 145 Hz of $^{1}J_{C\text{–}H}$, 1-s recycle delay, and 32 transients. The HMBC experiment is carried out with 128 data points in the indirect dimension and 4096 data points in F2 dimension, 8 Hz of $J_{C\text{–}H}$ coupling, 1-s recycle delay, and 32–128 transients accumulated.
4. The spectra are processed using Bruker Topspin software or available commercial software.

4 Notes

1. It is necessary to quantify the carbohydrate contents in the enzyme blank because some commercial enzymes contain carbohydrates that are stabilized with glucose [24].
2. Both electrochemical and RI detectors are available for carbohydrates analysis. Depending on the LC system, different detectors and columns can be applied [25, 26].
3. For the analysis of some target compounds, UV detector is needed [27].
4. Other GC columns are also available with different operating conditions.
5. Spectrum of the internal standard should be separated from other products spectra. The concentration of internal standard in each sample should be the same.
6. The suspension sample can also be stored at –20 °C for extended time until NMR analysis.

7. If the sample volume is less than 0.4 mL, then a 5 mm Shigemi NMR tube is used. Deuterated 2,2-dimethyl-2-silapentane-5-sulfonate (DSS-d_6, sodium salt; Sigma-Aldrich, St. Louis, MO) in D_2O (1.25 mM) can be also used as the locking solvent.
8. While standard probe is suitable for the experiments, sensitivity of detection dramatically improved with the cryoprobes. The cryoprobes are more expensive and provide two to four times better sensitivity than standard probes.
9. A relaxation reagent can be added to shorten the delay time.

Acknowledgments

This manuscript has been authored by UT-Battelle, LLC, under Contract No. DE-AC05-00OR22725 with the U.S. Department of Energy. This study was supported and performed as part of the BioEnergy Science Center (BESC) and the Center for Bioenergy Innovation (CBI). The BESC and CBI are U.S Department of Energy Bioenergy Research Centers supported by the Office of Biological and Environmental Research in the DOE Office of Science. The views and opinions of the authors expressed herein do not necessarily state or reflect those of the U. S. Government or any agency thereof. Neither the U. S. Government nor any agency thereof, nor any of their employees, makes any warranty, expressed or implied, or assumes any legal liability or responsibility for the accuracy, completeness, or usefulness of any information, apparatus, product, or process disclosed, or represents that its use would not infringe privately owned rights.

References

1. DuBois M, Gilles KA, Hamilton JK, Rebers P, Smith F (1956) Colorimetric method for determination of sugars and related substances. Anal Chem 28(3):350–356
2. Masuko T, Minami A, Iwasaki N, Majima T, Nishimura S-I, Lee YC (2005) Carbohydrate analysis by a phenol–sulfuric acid method in microplate format. Anal Biochem 339 (1):69–72
3. Ohemeng-Ntiamoah J, Datta T (2018) Evaluating analytical methods for the characterization of lipids, proteins and carbohydrates in organic substrates for anaerobic co-digestion. Bioresour Technol 247:697–704
4. Markou G, Angelidaki I, Nerantzis E, Georgakakis D (2013) Bioethanol production by carbohydrate-enriched biomass of Arthrospira (Spirulina) platensis. Energies 6 (8):3937–3950
5. Irick T, West K, Brownell H, Schwald W, Saddler J (1988) Comparison of colorimetric and HPLC techniques for quantitating the carbohydrate components of steam-treated wood. Appl Biochem Biotechnol 17(1-3):137–149
6. Schwald W, Chan M, Breuil C, Saddler J (1988) Comparison of HPLC and colorimetric methods for measuring cellulolytic activity. Appl Microbiol Biotechnol 28(4):398–403
7. Thangavelu SK, Ahmed AS, Ani FN (2016) Review on bioethanol as alternative fuel for spark ignition engines. Renew Sust Energ Rev 56:820–835
8. Ezeji T, Qureshi N, Blaschek HP (2007) Butanol production from agricultural residues: impact of degradation products on Clostridium beijerinckii growth and butanol fermentation. Biotechnol Bioeng 97(6):1460–1469

9. Thang VH, Kanda K, Kobayashi G (2010) Production of acetone–butanol–ethanol (ABE) in direct fermentation of cassava by *Clostridium saccharoperbutylacetonicum* N1-4. Appl Biochem Biotechnol 161(1–8):157–170
10. Tirado-Acevedo O, Chinn MS, Grunden AM (2010) Production of biofuels from synthesis gas using microbial catalysts. In: Advances in applied microbiology, vol 70. Elsevier, Amsterdam, pp 57–92
11. Lee SY, Park JH, Jang SH, Nielsen LK, Kim J, Jung KS (2008) Fermentative butanol production by clostridia. Biotechnol Bioeng 101 (2):209–228
12. Sreekumar S, Baer ZC, Pazhamalai A, Gunbas G, Grippo A, Blanch HW, Clark DS, Toste FD (2015) Production of an acetone-butanol-ethanol mixture from clostridium acetobutylicum and its conversion to high-value biofuels. Nat Protoc 10(3):528
13. Pi M-A, Simakova IL, Salmi T, Murzin DY (2013) Production of lactic acid/lactates from biomass and their catalytic transformations to commodities. Chem Rev 114(3):1909–1971
14. John RP, Nampoothiri KM, Pandey A (2007) Fermentative production of lactic acid from biomass: an overview on process developments and future perspectives. Appl Microbiol Biotechnol 74(3):524–534
15. Stoklosa R, Johnston D, Nghiem N (2018) Utilization of sweet sorghum juice for the production of Astaxanthin as a biorefinery co-product by Phaffia rhodozyma. ACS Sustain Chem Eng 6
16. Nghiem NP, Kim TH, Yoo CG, Hicks KB (2013) Enzymatic fractionation of SAA-pretreated barley straw for production of fuel ethanol and astaxanthin as a value-added co-product. Appl Biochem Biotechnol 171 (2):341–351
17. Ambati RR, Phang S-M, Ravi S, Aswathanarayana RG (2014) Astaxanthin: sources, extraction, stability, biological activities and its commercial applications—a review. Mar Drugs 12(1):128–152
18. Fahy E, Cotter D, Sud M, Subramaniam S (2011) Lipid classification, structures and tools. Biochim Biophys Acta 1811 (11):637–647
19. Le RK, Das P, Mahan KM, Anderson SA, Wells T, Yuan JS, Ragauskas AJ (2017) Utilization of simultaneous saccharification and fermentation residues as feedstock for lipid accumulation in Rhodococcus opacus. AMB Express 7(1):185
20. Kumar M, Saini S, Gayen K (2014) Acetone-butanol-ethanol fermentation analysis using only high performance liquid chromatography. Anal Methods 6(3):774–781
21. Oshiro M, Hanada K, Tashiro Y, Sonomoto K (2010) Efficient conversion of lactic acid to butanol with pH-stat continuous lactic acid and glucose feeding method by *Clostridium saccharoperbutylacetonicum*. Appl Microbiol Biotechnol 87(3):1177–1185
22. Swidah R, Wang H, Reid P, Ahmed H, Pisanelli A, Persaud K, Grant C, Ashe M (2015) Butanol production in *S. cerevisiae* via a synthetic ABE pathway is enhanced by specific metabolic engineering and butanol resistance. Biotechnol Biofuels 8(1):97
23. Carrieri D, McNeely K, De Roo AC, Bennette N, Pelczer I, Dismukes GC (2009) Identification and quantification of water-soluble metabolites by cryoprobe-assisted nuclear magnetic resonance spectroscopy applied to microbial fermentation. Magn Reson Chem 47(S1):S138–S146
24. Resch M, Baker J, Decker S (2015) Low solids enzymatic saccharification of lignocellulosic biomass. In: Laboratory analytical procedure (LAP) (NREL/TP-5100-63351). National Renewable Energy Laboratory, Golden, CO
25. Harde S, Wang Z, Horne M, Zhu J, Pan X (2016) Microbial lipid production from SPORL-pretreated Douglas fir by *Mortierella isabellina*. Fuel 175:64–74
26. Kim TH, Yoo CG, Lamsal B (2013) Front-end recovery of protein from lignocellulosic biomass and its effects on chemical pretreatment and enzymatic saccharification. Bioprocess Biosyst Eng 36(6):687–694
27. Nichols NN, Sharma LN, Mowery RA, Chambliss CK, Van Walsum GP, Dien BS, Iten LB (2008) Fungal metabolism of fermentation inhibitors present in corn Stover dilute acid hydrolysate. Enzym Microb Technol 42 (7):624–630

Chapter 10

Crystallography of Metabolic Enzymes

Markus Alahuhta, Michael E. Himmel, Yannick J. Bomble, and Vladimir V. Lunin

Abstract

The metabolic enzymes like any enzymes generally display globular architecture where secondary structure elements and interactions between them preserve the spatial organization of the protein. A typical enzyme features a well-defined active site, designed for selective binding of the reaction substrate and facilitating a chemical reaction converting the substrate into a product. While many chemical reactions could be facilitated using only the functional groups that are found in proteins, the large percentage or intracellular reactions require use of cofactors, varying from single metal ions to relatively large molecules like numerous coenzymes, nucleotides and their derivatives, dinucleotides or hemes. Quite often these large cofactors become important not only for the catalytic function of the enzyme but also for the structural stability of it, as those are buried deep in the enzyme.

Key words X-ray diffraction, Metabolic enzymes, Cofactors, Metabolic channels

1 Introduction

Allosteric regulation is the regulation of an enzyme by an effector molecule binding at a site that is different from the active site. That effector molecule could be a product of a chemical reaction catalyzed by an enzyme or a downstream product from the metabolic pathway or an upstream substrate from such pathway. The regulation occurs by conformational change in an enzyme caused by that allosteric binding resulting in active site becoming accessible or inaccessible or switching its ability to perform the reaction catalysis via changes in its geometry.

Metabolic enzymes are often known to form so-called metabolic channels—multienzyme complexes where several enzymes performing consecutive steps in a pathway are arranged to keep their active sites in close proximity [1]. That way the reaction intermediates does not have to travel far between conversion steps. In a crowded cell environment diffusion rates are diminished three- to four-fold even for small molecules [2], and more so for the

Michael E. Himmel and Yannick J. Bomble (eds.), *Metabolic Pathway Engineering*, Methods in Molecular Biology, vol. 2096,
https://doi.org/10.1007/978-1-0716-0195-2_10,

macromolecules [3]. That results in local concentrations of the intermediates being high at the metabolic channel while total cellular levels of these chemicals could be kept low. This, in turn, helps with an overall efficiency of the pathway on one hand and is important for cell well-being when such intermediates are toxic on another [4].

The co-localization could be achieved through fusion, compartmentalization, or use of scaffold. The first method, fusion, implies that the interactions between the different enzymes in such complex (which is more common for a primary metabolism) are extensive and might not only keep enzymes together but also could affect the stability and even the conformation of the "metabolic channel" members. Whenever a single enzyme from such complex is expressed, it could be unstable in absence of its binding partners.

Another feature of metabolic enzymes is their very high turnover rate compared to structural proteins or secreted proteins. While half-life of collagen and eye lens crystallin could be measured in decades, half-life of secreted enzymes could be often measured in days (cellulases), the average life time of many metabolic enzymes is counted in minutes [5, 6] as the proteins are synthesized on demand and are quickly recycled when no longer needed. That means that metabolic enzymes were not evolved for an extreme stability and resistance to proteolytic cleavage, or it even could be said that in general those enzymes with high turnover rate are quite susceptible to the proteolysis.

One more consideration is that metabolic enzymes are designed to work at very specific conditions with regard to temperature and chemical environment (including pH and protection from oxidation) as these parameters are much better controlled inside the cell. Secreted enzymes, on the other hand, are designed to withstand much harsher conditions outside of the cell where physical and chemical parameters could vary in much greater range.

Therefore, we can assume that, in general, metabolic enzymes are less stable than secreted proteins and require fine-tuning of chemical conditions such as pH, presence or absence of various metal ions and small molecules, and more protection from oxidation and proteolytic agents when overexpressed in large quantities for structural characterization.

2 Importance of Expression, Purification, and Modifications

The importance of protein expression and proper purification is often underestimated in the process of protein structure determination. A homogeneous solution of intact protein with the same modifications is needed for all structure determination methods. A robust purification regime and/or *C*-terminal affinity tags can guarantee the purification of only fully translated gene products. However, post-translational modifications such as protein maturation, regulation, and covalent modifications in general are much harder to control. An obvious solution is to express the target protein in an organism which is not able to perform these actions. *E. coli* often is the first choice due to its well-established methods, fast growth rate, and lack of many post-translational modification systems compared to other organisms [7–9].

Glycosylation is an example of a post-translational modification that cannot be done by *E. coli* without metabolic engineering of the *E. coli* expression host [10]. Often this lack of post-translational modification systems in *E. coli* turns into a problem when producing eukaryotic proteins where modifications such as glycosylation are necessary for solubility. Multiple eukaryotic and cell-free expression systems have been developed to address this issue [8, 11, 12]. Commonly used systems are yeast [13], wheat germ [14], insect cells [15], rabbit reticulocytes [16], tumor HeLa cells [17], and hybridoma [18]. All host organisms have their disadvantages and advantages. Some are harder to work with or have low yields making it necessary to understand and choose the expression host according to one's specific post-translational modification needs.

In many cases, it is not possible to find a perfect expression system or the modification in question cannot be introduced to all molecules. For these cases, thorough purification and verification of purity is needed. Incomplete mRNA translation or protease action can lead to partial proteins that are usually, but not always or completely, removed by the host cell. These incomplete proteins can still be soluble and partially folded. If they are not removed from the protein solution before crystallization experiment, they can lower the effective concentration, and in extreme cases, block crystal growth by binding the crystal surface and by not providing further binding sites for more molecules due to their missing parts. If the missing parts are large enough, size exclusion chromatography can help.

Affinity purification methods with *C*-terminal His-tags [19] or Strep-tags [20] can be used to effectively remove incomplete translation products. If translation is terminated early, the tag won't be translated and the resulting incomplete protein will not bind the affinity resin. It should be noted that affinity tag purification methods should not be used as the only purification step. When using these methods, it does not matter if the protein is damaged, incomplete (when using N-terminal tags), unfolded, or inactive if the tag is present and the protein part in question is soluble. Affinity tag purification steps should always be followed by a complementary affinity chromatography method (ion exchange, hydrophobic interaction) and/or size exclusion chromatography to remove these impurities. Proper purification is particularly necessary before crystallizing proteins with covalent modifications. These modifications are rarely uniform and typically affect surface properties. This makes affinity chromatographic steps such as hydrophobic interaction and ion exchange very effective at separating differently modified populations.

3 Practical Considerations on How to Purify Tagged Proteins for Crystallization

1. Be aware of solubility and stability of your protein before purification (salt concentration), solubilizing chemicals (detergents, non-detergent sulfobetaines), cofactors, metals, reducing agents and oxygen scavengers.
2. Prepare lysis buffer with possible stabilizing and solubilizing agents and do lysis using your preferred method.
   (a) E.g., 50 mM Tris pH 7.5, 100 mM NaCl.
3. Perform affinity purification according to manufacturer's instructions.
   (a) For Nickel-NTA: remove cell debris by centrifugation at 15,000 × *g* for 15 min; dilute the supernatant with equal volume of binding buffer (50 mM Tris pH 7.5, 100 mM NaCl, and 20 mM imidazole); load into a 5 mL HisTrap column (GE Healthcare, Piscataway, NJ); wash the unbound proteins from the column using five column volumes of binding buffer; elute with elution buffer (50 mM Tris pH 7.5, 100 mM NaCl, and 250 mM imidazole).
4. Verify homogeneity with SDS-PAGE, mass spectrometry, and light scattering.
5. Run size exclusion chromatography.
   (a) For example, HiLoad Superdex 75 (16/60) column (GE Healthcare, Piscataway, NJ) in 20 mM Tris pH 7.5, 100 mM NaCl.
6. Verify homogeneity with SDS-PAGE, mass spectrometry, light scattering.
7. If not pure, run hydrophobic interaction and/or ion exchange chromatography.
   (a) Hydrophobic interaction: change the sample to 50 mM Bis–Tris pH 6.5, 2 M NH_4SO_4; load onto a Tricorn 10/100 column (GE Healthcare, Piscataway, NJ) packed with Source 15Phe Hydrophobic Interaction Chromatography medium (GE Healthcare); elute using a 25-column volume linear gradient of 20 mM Bis–Tris, pH 6.5, and 2 to 0 M NH_4SO_4.
   (b) Anion exchange: desalt (dialyze) into 20 mM Bis–Tris pH 6.5; load onto a Tricorn 10/100 column packed

with Source 15Q anion-exchange medium (GE Healthcare); elute with a 25-column volume linear gradient of 20 mM Bis–Tris pH 6.5 and 0 to 1 M NaCl.

8. Repeat the size exclusion chromatography step or dialyze to a buffer concentration of 20 mM and other conditions suitable for your protein.
   (a) It is important to have a low enough buffer concentration to allow crystallization screen buffers (at 100 mM) to change the pH. Up to 50 mM can be used, but 20 mM is preferred.

4 Crystallization Techniques

The history of protein crystallization started by chance during the nineteenth century with stable extracellular proteins such as hemoglobin [21, 22]. More practical crystallization methods during the twentieth century allowed structure determination of other proteins including metabolic enzymes [23]. Finally, high throughput crystallization and Structural Genomics Consortiums exploded the number of available structures [24, 25]. Unfortunately, this means that the structures of most of the easy-to-crystallize proteins have been done. Due to advancements in metabolic engineering, the focus of structural biology is increasingly on the short-lived metabolic enzymes that are subject to regulation, have cofactors, and often are modified by post-translational modifications. This means that the precise control of protein purity and homogeneity must be coupled with crystallization screens containing different substrates, inhibitors, cofactors, regulators, and metals for successful crystallization experiments. Proteins, in general, have a wide array of properties that affect crystallization. They differ in size, shape, mobility, stability, and preference for the overall conditions they are subjected to (small molecule composition, pH, ionic strength, temperature, pressure). These properties are often dictated by the protein's natural conditions and cellular environments. This helps to define starting conditions for crystallization experiments.

In simplified terms, protein crystallization is composed of two steps: (1) identification of conditions that yield crystalline material and (2) the incremental optimization of these conditions (varying pH, precipitant, etc.) to produce crystals that can be optimally used for X-ray diffraction and structure determination [23]. The first step requires understanding of the target proteins natural function (substrates/inhibitors, cofactors, regulation) and screening for crystallization conditions. Generally, screening can be systematic or random and usually vary according to protein concentration, precipitant type and concentration, pH, and temperature. In both cases, conditions can be affected by prior knowledge of target proteins function and natural environment, that is, screens are done with or without substrates/inhibitors, cofactors, etc. While the systematic approach is thorough, it usually requires more protein and can be unfeasible making it a better fit for the optimization step. In contrast, the random approach needs less protein but may not perfectly capture all conditions. In practice, random screens using sparse-matrix sampling in combination with robotics [26–28] are more commonly used and readily available commercially.

Finding a hit from the initial screens starts the optimization part. A hit can be a crystal good enough for X-ray diffraction experiments, a micro crystal, cluster of crystals, or just crystalline material. In all cases, it is important to verify that the crystalline material is protein and not a small molecule from the crystallization conditions. Most reliable way to test this is to mount the crystal and get an X-ray diffraction pattern, but other ways beyond the scope of this chapter are available (for a great recap of available methods, *see*: https://www.hamptonresearch.com/documents/product/hr007641_cg101_salt_or_protein_crystals.pdf). After the crystalline material has been verified to be protein an optimization screen should be prepared, where the pH and precipitant type and concentration are systematically varied [29]. A popular method that can be used in combination with optimization screens is seeding [30–32]. Microseeding uses microcrystals and macroseeding starts with larger macrocrystals. Both methods aim to slowly and controllably grow large high-quality crystals out of existing smaller or lower quality ones in fresh crystallization solutions.

Vapor diffusion is the most commonly used crystallization technique and will be briefly described here. Other methods used but not discussed here are microbatch, microdialysis, and free-interface diffusion. All techniques have their uses, and their benefits have been extensively discussed and compared in the literature [33, 34]. In vapor diffusion, the crystallization drop with a mixture of sample and reagent is placed next to a liquid reservoir of reagent in a sealed chamber. The crystallization drop can be hanging from a siliconized cover plate over the reservoir solution ("hanging drop" vapor diffusion) or sitting next to it ("sitting drop" vapor diffusion). The "sitting drop" method is commonly used with crystallization robots. The starting concentration of the precipitant is lower in the crystallization drop, because it contains a mixture of both the protein sample and the reservoir solution. Water and other volatile liquids will slowly evaporate and reach an equilibrium between the crystallization drop and the reservoir solution. This leads to gradually increasing concentration of precipitant and protein in the crystallization drop. If conditions are right for nucleation, a crystal will start growing in the increasingly higher protein and precipitant concentrations. The speed of this reaction can be controlled by drop and reservoir volume and ratio; by varying precipitant type and concentration; and by changing other environmental parameters like temperature, pressure, chamber size, vibration, and sound.

5 Crystallizing Proteins

1. Properly purify protein and verify homogeneity.
   - (a) Affinity tag purification followed by size exclusion chromatography.
   - (b) SDS-PAGE, mass spectrometry, light scattering.
2. Setup an initial crystallization screen or use a commercial concentration determination kit to determine correct crystallization screen protein concentration.
   - (a) 10 mg/mL is a good concentration to start with for most medium-sized (20–80 kDa) proteins.
3. Setup crystallization screens according to manufactures instructions with and without known substrates, inhibitors, cofactors, regulators, and metals at different temperatures and protein concentrations.
   - (a) Several commercial screens are available. The most commonly used is Crystal Screen by Hampton Research (Aliso Viejo, CA).
   - (b) Start with the most likely combinations such as required cofactors, metals, and possibly inhibitors.
4. Inspect plates using a microscope in regular intervals.
   - (a) Once a day, every other day, once a week.
5. Optimize hits using optimization screens, microseeding, etc.
   - (a) A simple optimization screen varies pH and precipitant. For example, if hit was 0.1 M citric acid pH 5.5 and 1.5 M ammonium sulfate, a simple optimization screen would be 0.1 M citric pH 5–6 with 1.0–2.0 M ammonium sulphate.
6. Be patient and do not give up. Sometimes crystallization takes time and conditions change over time (diffusion trough plastic, etc.)

6 BDH Example

6.1 Background

2,3-Butanediol (2,3-BD) is considered to be a possible biofuel as well as a valuable precursor for a further chemical conversion. 2,3-BD is a byproduct of a sugar metabolism in some microorganisms. Starting from pyruvate, it takes three enzymes to obtain 2,3-BD. First, acetolactate synthase (ALS) converts two molecules of pyruvate into acetolactate (AL). Then acetolactate decarboxylase (ALDC) converts AL into acetoin (AC) which is further converted into 2,3-BD by butanediol dehydrogenase (BDH). BDH enzymes require NADH or NADPH as a reducing agent, and the reaction is reversible with the direction of the reaction governed by pH [35, 36].

$$AC + NADH + H^+ \leftrightarrow 2,3 - BD + NAD^+$$

Based on chiral preferences for the substrate, BDH enzymes could be classified as R-AC or S-AC dependent, and depending on a chirality of the chiral center introduced by the enzyme, they could be classified as S-acting or R-acting.

While the preference for the R-AC or S-AC is imprinted in the geometry of the substrate-binding pocket, S-acting and R-acting BDHs belong to different families and possess different architectures. To the date, no structures of R-acting BDHs were reported.

S-acting BDHs belong to the short-chain dehydrogenase/reductase (SDR) family. The substrates of enzymes belonging to that family vary greatly in size and include glucose, alcohols, and steroids [37]. This family is well studied and many members of it are characterized structurally including several S-acting BDHs. Here we will describe the structural studies of two S-acting BDHs, R-AC dependent S-acting BDH from *Klebsiella pneumoniae* (PDB ID 1GEG, deposited in 2001) and S-AC dependent S-acting BDH from *Corynebacterium glutamicum* (PDB ID 3A28, deposited in 2010) reported by Masato Otagiri and colleagues [38, 39].

According to the authors [40], both enzymes were cloned and overexpressed in *E. coli*. No affinity tags (e.g., 6xHis tags) were used. Protein purification was carried out using ammonium sulfate precipitation followed by subsequent gel filtration chromatography. While use of affinity tags is a widely used technique, one should always be careful designing the construct as those tags could interfere with protein–protein interactions within a multimer or within a multienzyme complex. In the case of S-acting BDHs which are found to be homotetramers (or more precise, dimers of dimers), use of *C*-terminal tags would disrupt formation of the multimer, and therefore greatly affect enzyme activity and stability. We found that use of the *N*-terminal His-tag for a S-acting BDH from a different species did not interfere with the tetramerization and protein crystallization process.

The crystals for the proteins of interest were obtained using hanging drop vapor diffusion technique with polyethylene glycol (PEG) 4000 as a precipitating agent. Both enzymes were co-crystallized with NAD^+, and at the first attempt, authors tried to trap a meso-2,3-BD in the active site of the enzyme so that compound was also added to the crystallization solution. It turned out that another small molecule was trapped in the active site instead. β-mercaptoethanol (BME) was used in high concentration as a part of a cryoprotection solution, and upon analysis of the refined structure, it was decided that it was BME, not 2,3-BD trapped in the active site.

We re-refined both structures using experimental data that were deposited alongside the protein models. The benefits of the modern crystallographic software were clearly displayed as we were able to significantly improve the quality of the models following the details that became visible in the electron density maps along with corrections of the obvious modeling errors. These improvements were reflected in the R/R_{free} factors declined from 0.193/0.209 to 0.120/0.162 for 1GEG and from 0.193/0.240 to 0.152/0.214 for 3A28. Improvements to the models included corrected and much more complete water structure, alternative conformations of the sidechains, corrected sidechains positions (including Asn/-His/Gln flips). The overall main chain trace was already correct in the deposited models and large cofactor NAD^+ molecules were placed correctly. Initial placement of inhibitor BME molecules seems to be correct in the case of 1GEG and is questionable in 3A28. We removed BME molecules placed in the active sites of 3A28 entry and analyzed the resulting omit electron density. We found it unsupportive of BME placement, rather it was found to be a mix of water molecules and small solvent molecules like ethylene glycol and glycerol that are widely used for the cryoprotection of the protein crystals. No strong electron density corresponding to a sulfur atom was found in any of the eight active sites of the 3A28 entry. A mix of ethylene glycol and glycerol molecules was modeled instead of BME.

6.2 Overall Structure

As it was mentioned earlier, an S-acting BDH forms a homotetramer (to be more precise, a dimer of dimers, Fig. 1a). A single subunit of the BDH possess a dinucleotide-binding Rossmann fold with 7-stranded parallel β-sheet flanked by three long α-helices on each side of it. Two more short α-helices in a region 186-221 for 1GEG (188-223 for 3A28) form a small lobe on top of a core structure creating a deep cleft in which the cofactor is bound and the active site is located (Fig. 1b).

6.3 NAD^+ Binding

In *K. pneumoniae*, BDH NAD^+ cofactor binding is facilitated through a network of hydrogen bonds that includes a water-mediated interaction as well as some Van der Waals interactions. The amino acid residues involved in NAD^+ binding are shown and labeled in the Fig. 2. The GXXXGX sequence between the first β-strand and first α-helix is a 'signature' of the dinucleotide-binding motif in SDR enzymes. The sidechain of the Asp33 that is involved in H-bonds with the 2′- and 3′-hydroxyl groups of the adenine ribose is common in the SDR enzymes that exclusively bind

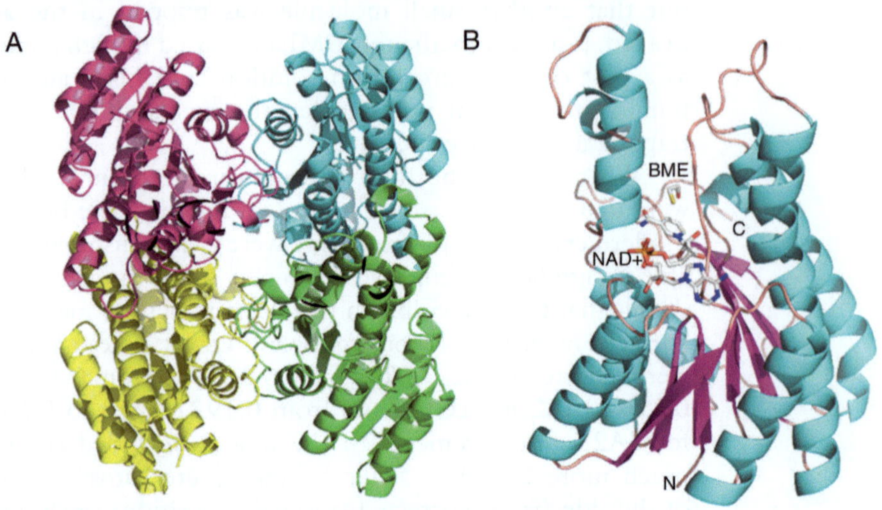

Fig. 1 (**a**) The homotetramer of *K. pneumoniae* BDH (PDB ID 1GEG) in cartoon representation colored by molecule. (**b**) The monomer of *K. pneumoniae* BDH in cartoon representation colored by secondary structure. Cofactor NAD^+ and substrate analog BME are shown in sticks representation

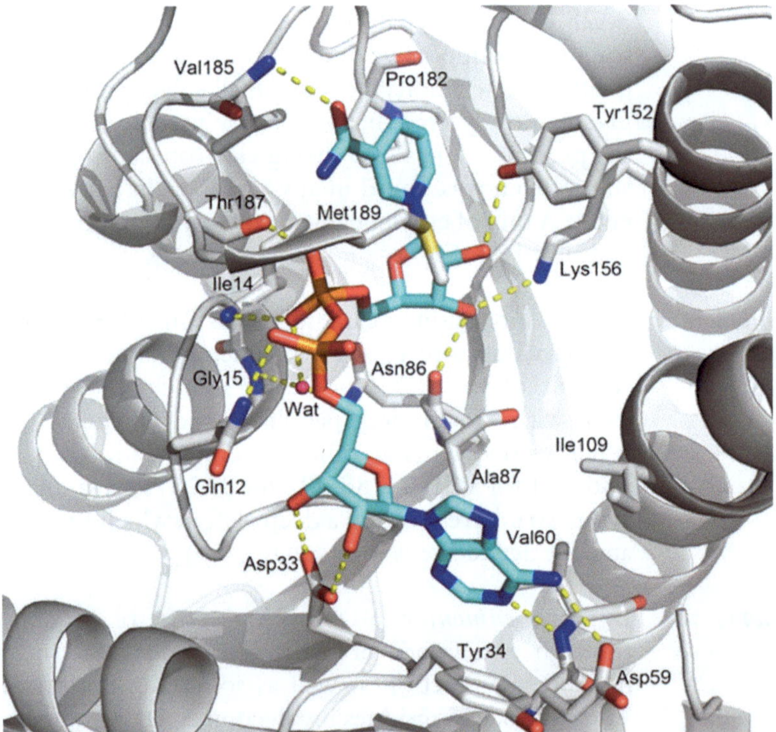

Fig. 2 Cofactor binding site in *K. pneumoniae* BDH. Amino acid residues involved in H-bonds or VDW interactions with the NAD^+ are shown in sticks with grey carbons. NAD^+ molecule is shown in sticks with cyan carbons. Hydrogen bonds are shown as yellow dashed lines

$NADH/NAD^+$. A glycine residue at this position allows enzyme to accommodate $NADPH/NADP^+$ instead.

The NAD^+ binding site in *C. glutamicum* is virtually identical to that of *K. pneumoniae* with Tyr34 residue replaced by leucine.

6.4 Substrate Analog Binding

The cavities where the substrate analog molecules were found in both *K. pneumoniae* and *C. glutamicum* BDHs are mostly formed by hydrophobic sidechains with few strategically placed H-bond donors. In case of *K. pneumoniae* the walls of the pocket are made by Ala90, Ser139, Gln140, Ala141, Asn146, Leu149, Tyr152, Pro182, Gly183, Ile184, Met189, Trp190, Ile193, Phe212, and Met253. Ser139, Gln140, Tyr152, and Gly183 could act as the hydrogen bond partners with the substrate atoms. Ser139 and Tyr152 are also the part of catalytic tetrad (Asn110, Lys156, Ser139, Try152). The bottom of the pocket is capped with the nicotinamide part of the NAD^+ and the opening to the solvent is gated by Phe212, Gln140, and Asn146 sidechains. The sidechain of the Trp190 is held in place via the H-bond between Trp190 Nε and Thr209 Oγ. We cannot be sure that the position of BME is completely representative of the placement of the acetoin/2,3-BD in that pocket because of the presence of a larger sulfur atom, C–S bond being longer than C–O and the differences between sulfur and oxygen atoms with regard to participation in hydrogen bonds. The S atom of the BME sits 2.3 Å away from the C4N of the NAD^+ and 3.2 Å from the Tyr152 hydroxyl, while C2 of the BME is 3.3 Å away from the C4N atom of the NAD^+. The O atom of the BME could be engaged in the H-bonds with Nε2 of the Gln140 (3.0 Å), Oγ of the Ser139 (2.8 Å), or the main chain carbonyl of the Gly183 (2.6 Å).

In the *C. glutamicum* BDH, the substrate-binding pocket is formed by Ala92, Ser141, Ile142, Ala143, Phe148, Leu151, Pro184, Gly185, Ile186, Met191, Trp192, Ile195, Tyr214, and Met255. Ser141, Tyr154, Gly185, and Trp192 could provide possible H-bonds to the substrate/product. As with the other BDH, Ser141 and Tyr154 belong to the catalytic tetrad (Asn112, Lys158, Ser141, Tyr154). The bottom of the pocket is capped with the nicotinamide part of the NAD^+ and the opening to the solvent is gated by Phe148 and Tyr214 sidechains. The main differences from *K. pneumoniae* pocket are the loss of H-bond donor Gln140 that is replaced with Ile142 in *C. glutamicum*, corresponding change of Asn146 to Phe148 and the flip of the sidechain of Trp190 (Trp192 in *C. glutamicum*) which now provides a H-bond to the hydroxyl of the substrate analog. The flip of the Trp192 sidechain is caused by Thr209 of *K. pneumoniae* now replaced by Phe211 in *C. glutamicum*, removing the H-bond partner and taking up the space previously occupied by Trp sidechain.

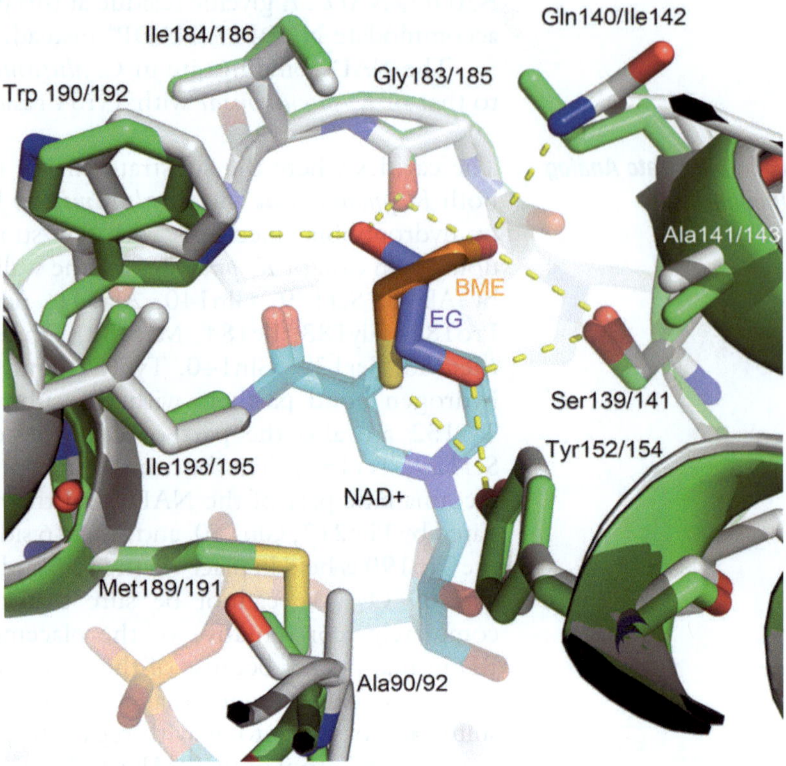

Fig. 3 Superimposed substrate binding pockets from *K. pneumoniae* (grey carbons for the protein atoms and orange carbons for the BME substrate analog) and *C. glutamicum* (green carbons for protein and purple for the EG substrate analog) BDHs. NAD^+ cofactor is shown in sticks with cyan carbons. The most important differences in the pocket organization are the flip of the Trp190/192 sidechain and Gln140 of the *K. pneumoniae* replaced by Ile142 in *C. glutamicum*. Possible hydrogen bonds are shown as yellow dashed lines

We believe that after replacing the BME with the ethylene glycol and refining the structure, the position of the substrate analog is representative of that of the S-acetoin/2S,3S-BD (L-BD). The O1 hydroxyl is engaged in two H-bonds with Trp192 Nε (2.7 Å) and main chain carboxyl of Gly185 (2.7 Å). The O2 hydroxyl is making contacts with Tyr154 sidechain hydroxyl (2.5 Å) and Ser141 Oγ (2.6 Å). C1 and C2 atoms of the EG now represent C2 and C3 atoms of the acetoin/2,3-BD. The superimposed substrate binding pockets of two BDHs with the most important interactions between the protein and the substrate analog are shown in Fig. 3.

6.5 Stereospecificity

Based on ethylene glycol position in the *C. glutamicum* BDH structure, we can now model not only 2S,3S-BD in that active site but also a meso-BD in the active site of the *K. pneumoniae* BDH with the O2 and C2 and C3 carbon atoms of the 2,3-BD

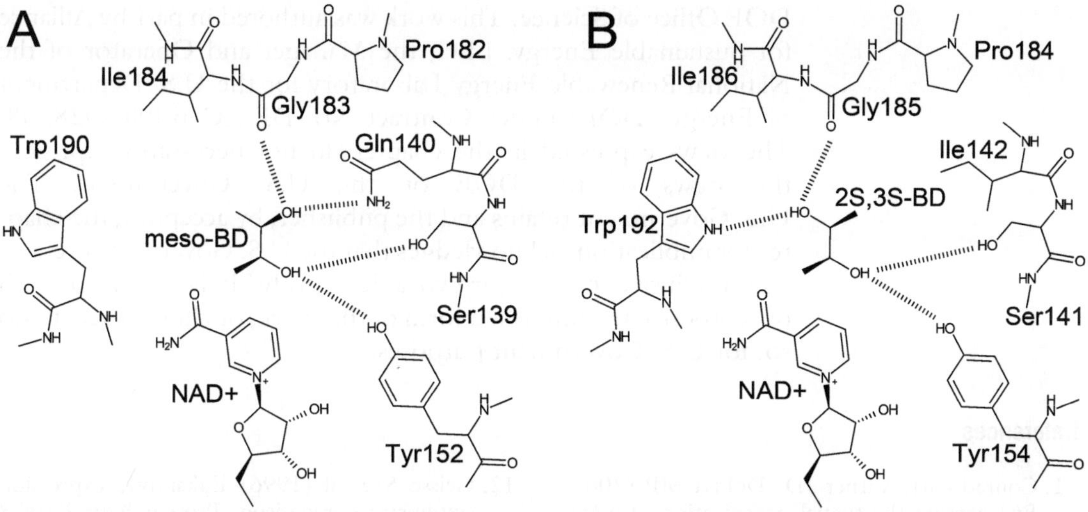

Fig. 4 (**a**) Binding of a meso-BD in *K. pneumoniae* BDH. (**b**) Binding of a 2S,3S-BD in *C. glutamicum* BDH

remaining in the same positions (Fig. 4). The difference in stereospecificity of these two BDHs is simply explained by the position of the H-bond partner suitable to bind the acetoin hydroxyl in mostly hydrophobic environment of the binding pocket. That partner is Nε or the Gln140 in *K. pneumoniae* BDH and Nε or the Trp192 in *C. glutamicum* BDH.

7 Lessons Learned

The X-ray crystallography is a valuable tool which allows us to study complex interactions between an enzyme, cofactor, and its substrate at the atomic level. At the same time, it is very important to carefully control what chemicals are present in the crystallization mix and thoroughly analyze the chemical identity of the molecules found in an active site of an enzyme (if any). Also, it must be understood that when a substrate analog is discovered in an active site, its position might be not representative of that of a real substrate and any modeling and suggestions for the catalytic mechanism should be carried out with a great deal of caution and skepticism.

Acknowledgments

Funding was provided by the BioEnergy Science Center (BESC) and the Center for Bioenergy Innovation (CBI), from the U.S. Department of Energy Bioenergy Research Centers supported by the Office of Biological and Environmental Research in the

DOE Office of Science. This work was authored in part by Alliance for Sustainable Energy, LLC, the Manager and Operator of the National Renewable Energy Laboratory for the U.S. Department of Energy (DOE) under Contract No. DE-AC36-08GO28308. The views expressed in the chapter do not necessarily represent the views of the DOE or the U.S. Government. The U.S. Government retains and the publisher, by accepting the chapter for publication, acknowledges that the U.S. Government retains a nonexclusive, paid-up, irrevocable, worldwide license to publish or reproduce the published form of this work, or allow others to do so, for U.S. Government purposes.

References

1. Conrado RJ, Varner JD, DeLisa MP (2008) Engineering the spatial organization of metabolic enzymes: mimicking nature's synergy. Curr Opin Biotechnol 19(5):492–499
2. Kao HP, Abney JR, Verkman AS (1993) Determinants of the translational mobility of a small solute in cell cytoplasm. J Cell Biol 120(1):175
3. Arrio-Dupont M et al (2000) Translational diffusion of globular proteins in the cytoplasm of cultured muscle cells. Biophys J 78 (2):901–907
4. Meynial Salles I et al (2007) Evolution of a *Saccharomyces cerevisiae* metabolic pathway in *Escherichia coli*. Metab Eng 9(2):152–159
5. Toyama BH, Hetzer MW (2012) Protein homeostasis: live long, won't prosper. Nat Rev Mol Cell Biol 14:55
6. Eden E et al (2011) Proteome half-life dynamics in living human cells. Science 331 (6018):764
7. Rosano GL, Ceccarelli EA (2014) Recombinant protein expression in *Escherichia coli*: advances and challenges. Front Microbiol 5:172
8. Tokmakov AA et al (2012) Multiple post-translational modifications affect heterologous protein synthesis. J Biol Chem 287 (32):27106–27116
9. Brown CW et al (2017) Large-scale analysis of post-translational modifications in *E. coli* under glucose-limiting conditions. BMC Genomics 18(1):301
10. Makino T, Skretas G, Georgiou G (2011) Strain engineering for improved expression of recombinant proteins in bacteria. Microb Cell Factories 10:32
11. Zemella A et al (2015) Cell-free protein synthesis: pros and cons of prokaryotic and eukaryotic systems. Chembiochem 16 (17):2420–2431
12. Geisse S et al (1996) Eukaryotic expression systems: a comparison. Protein Expr Purif 8 (3):271–282
13. Eckart MR, Bussineau CM (1996) Quality and authenticity of heterologous proteins synthesized in yeast. Curr Opin Biotechnol 7 (5):525–530
14. Madin K et al (2000) A highly efficient and robust cell-free protein synthesis system prepared from wheat embryos: plants apparently contain a suicide system directed at ribosomes. Proc Natl Acad Sci U S A 97 (2):559–564
15. Ezure T et al (2006) Cell-free protein synthesis system prepared from insect cells by freeze-thawing. Biotechnol Prog 22(6):1570–1577
16. Merrick WC, Barth-Baus D (2007) Use of reticulocyte lysates for mechanistic studies of eukaryotic translation initiation. Methods Enzymol 429:1–21
17. Mikami S et al (2006) An efficient mammalian cell-free translation system supplemented with translation factors. Protein Expr Purif 46 (2):348–357
18. Mikami S et al (2006) A hybridoma-based in vitro translation system that efficiently synthesizes glycoproteins. J Biotechnol 127 (1):65–78
19. Hochuli E et al (1988) Genetic approach to facilitate purification of recombinant proteins with a novel metal chelate adsorbent. Bio/-Technology 6:1321
20. Schmidt TG, Skerra A (2007) The strep-tag system for one-step purification and high-affinity detection or capturing of proteins. Nat Protoc 2(6):1528–1535
21. Giege R (2013) A historical perspective on protein crystallization from 1840 to the present day. FEBS J 280(24):6456–6497

22. Hünefeld FL (1840) Der Chemismus in der thierischen Organisation: physiologisch-chemische Untersuchungen der materiellen Veränderungen oder des Bildungslebens im thierischen Organismus, insbesondere des Blutbildungsprocesses, der Natur der Blutkörperchen und ihrer Kernchen: ein Beitrag zur Physiologie und Heilmittellehre, Leipzig
23. McPherson A, Gavira JA (2014) Introduction to protein crystallization. Acta Crystallogr F Struct Biol Commun 70(Pt 1):2–20
24. Kim Y et al (2011) High-throughput protein purification and quality assessment for crystallization. Methods 55(1):12–28
25. Structural Genomics Consortium et al (2008) Protein production and purification. Nat Methods 5(2):135–146
26. Chayen NE et al (1990) An automated system for micro-batch protein crystallization and screening. J Appl Crystallogr 23(4):297–302
27. Cox MJ, Weber PC (1987) Experiments with automated protein crystallization. J Appl Crystallogr 20(5):366–373
28. Jancarik J, Kim S-H (1991) Sparse matrix sampling: a screening method for crystallization of proteins. J Appl Crystallogr 24(4):409–411
29. Weber PC (1990) A protein crystallization strategy using automated grid searches on successively finer grids. Methods 1(1):31–37
30. Bergfors T (2003) Seeds to crystals. J Struct Biol 142(1):66–76
31. Thaller C et al (1985) Diffraction methods for biological macromolecules. Seed enlargement and repeated seeding. Methods Enzymol 114:132–135
32. D'Arcy A, Villard F, Marsh M (2007) An automated microseed matrix-screening method for protein crystallization. Acta Crystallogr D Biol Crystallogr 63(Pt 4):550–554
33. D'Arcy A et al (2003) The advantages of using a modified microbatch method for rapid screening of protein crystallization conditions. Acta Crystallogr D Biol Crystallogr 59 (Pt 2):396–399
34. Chayen NE (1998) Comparative studies of protein crystallization by vapour-diffusion and microbatch techniques. Acta Crystallogr D Biol Crystallogr 54(Pt 1):8–15
35. Zhang L et al (2014) A new NAD(H)-dependent meso-2,3-butanediol dehydrogenase from an industrially potential strain Serratia marcescens H30. Appl Microbiol Biotechnol 98(3):1175–1184
36. Zhang X et al (2014) Two-stage pH control strategy based on the pH preference of Acetoin Reductase regulates Acetoin and 2,3-butanediol distribution in Bacillus subtilis. PLoS One 9(3):e91187
37. Jörnvall H et al (1995) Short-chain dehydrogenases/reductases (SDR). Biochemistry 34 (18):6003–6013
38. Otagiri M et al (2001) Crystal structure of meso-2,3-butanediol dehydrogenase in a complex with NAD^+ and inhibitor Mercaptoethanol at 1.7 a resolution for understanding of chiral substrate recognition Mechanisms1. J Biochem 129(2):205–208
39. Otagiri M et al (2010) Structural basis for chiral substrate recognition by two 2,3-butanediol dehydrogenases. FEBS Lett 584(1):219–223
40. Ui S et al (1998) Cloning, expression and nucleotide sequence of the l-2,3-butanediol dehydrogenase gene from Brevibacterium saccharolyticum C-1012. J Ferment Bioeng 86 (3):290–295

Chapter 11

Measuring Metabolic Enzyme Performance

Amanda M. Williams-Rhaesa and Michael W. W. Adams

Abstract

Understanding the performance of key metabolic enzymes is critical to metabolic engineering. It is important to know the kinetic parameters of both native enzymes and heterologously expressed enzymes that play key roles in pathway performance (Zeldes et al., Front Microbiol 6:1209, 2015; Keller et al., Metab Eng 27:101–106, 2015). This step cannot be overlooked as gene expression is not always a good indicator of the production of fully active enzymes, especially those requiring cofactor assembly and processing (Zeldes et al., Front Microbiol 6:1209, 2015; Chandrayan et al., J Biol Chem 287:3257–3264, 2012; Basen et al., MBio 3:e00053–e00012, 2012). Additionally, knowing kinetic parameters and having accurate and reproducible assays allows for the use of powerful computational and in vitro pathway optimization tools that can inform metabolic engineering efforts that in turn can lead to improvements in pathway performance (Keller et al., Metab Eng 27:101–106, 2015; Copeland et al., Metab Eng 14:270–280, 2012). To take full advantage of these tools, understanding the roles of both enzymes directly involved in a pathway of interest, together with those in related pathways that may syphon off key intermediates, is ideal (Keller et al., Metab Eng 27:101–106, 2015; Thorgersen et al., Metab Eng 22:83–88; Lin et al., Metab Engi 31:44–52, 2015).

Key words Metabolic engineering, Enzyme kinetics, Metabolic enzymes

1 Introduction

The gold standard of biochemical analysis is the determination of kinetic parameters of a purified enzyme in vitro. These data help to confirm optimal activity for heterologously expressed enzymes as well as for over-expressed native enzymes [1, 3, 4]. Under these conditions, limitations in processing genes and cofactor production can lead to enzymes with lower than expected activity even when gene expression is high. Ensuring that overexpressed pathway genes result in the production of fully functional enzymes is critical to metabolic engineering efforts. These data also help to elucidate the likely rate-limiting steps in a pathway, which can often be optimized through altered enzyme stoichiometry. Finding the optimal ratios of pathway enzymes can be sought through genetic

Michael E. Himmel and Yannick J. Bomble (eds.), *Metabolic Pathway Engineering*, Methods in Molecular Biology, vol. 2096, https://doi.org/10.1007/978-1-0716-0195-2_11,

means based on the results from in vitro assays using differing ratios of purified enzymes [2, 5–7].

There are several challenges to determining the kinetic parameters of key metabolic enzymes, including difficulty in protein purification and the limitations of the assays due to substrate stability. Obtaining a pure protein is desired for most kinetic analyses as this ensures that the properties of the protein of interest are not changed by contaminating proteins. Purification can be challenging as some enzymes may be unstable during purification or may lose activity due to loss of a cofactor or of other post-translational modifications, such as phosphorylation or glycosylation. However, purification is not always necessary and the use of impure enzymes or even cell-free extracts can overcome some of these limitations and provide some basic kinetic information. Substrate stability might also be an issue as some common metabolic intermediates are not very stable and have to be frozen and stored and used as soon as solutions are prepared. In these cases, linked assays are often the solution where the unstable intermediate is generated as it is needed, however, this may not be an option if the regeneration enzymes cannot be obtain or be purified.

Key metabolic enzymes are not only those involved in primary carbon metabolism but can also be involved in electron transfer or redox reactions. In the following, we describe the assays for two key types of redox enzyme in anaerobic metabolism: pyruvate ferredoxin oxidoreductase (POR) and hydrogenase (H_2ase), both bifurcating and nonbifurcating. POR catalyzes the pyruvate-dependent reduction of the iron-sulfur-containing redox protein ferredoxin. POR was selected as it catalyzes a critical reaction in a variety of anaerobic microorganisms by committing carbon to the production of acetyl-CoA, a key primary pathway intermediate [8]. H_2ases are also very common in anaerobic microorganisms and are pivotal to redox recycling to regenerate either oxidized ferredoxin or NAD, or both, to allow metabolic flux to continue. Redox recycling is often overlooked in metabolic engineering efforts [9].

The assays for both POR and H_2ase are very specific for these enzymes and are therefore effective using cell-free extracts as well as purified proteins. However, these enzymes are oxygen-sensitive, and so the assays are carried out under strictly anaerobic conditions. This means that all solutions must be degassed to remove oxygen and be maintained under a positive pressure of an inert gas. In the following, we use argon (Ar) as it is heavier than air, but dinitrogen (N_2) can also be used. Both the POR and H2ase assays also use methyl viologen (MV) as the artificial electron carrier. However, the assay for a bifurcating H_2ase by definition requires three (rather than two) specific substrates, and an artificial electron carrier cannot be used. One of those substrates for the bifurcating H_2ase is reduced ferredoxin and this is generated using the POR reaction.

2 Equipment

Gas chromatograph

Spectrophotometer with temperature control

Heated and shaking water bath

Cuvettes and stoppers

Anaerobic vials, butyl stoppers, crimp seals

Hamilton syringes

3 Abbreviations

POR	Pyruvate oxidoreductase
H_2ase	Hydrogenase
MV	Methyl viologen
TPP	Thiamine pyrophosphate
DT	Sodium dithionite
Fd	Ferredoxin

4 Preparation of Anaerobic Buffers, Reagents, and Cuvettes

1. Degas the appropriate buffer under vacuum while stirring for 20–30 min three times alternating with argon. End with argon in the headspace.
2. For powdered components, weigh appropriate amount of reagent into a 5 mL crimp seal vial, seal, and degas vial (three cycles of 1–2 min of alternating gas and vacuum, end under argon). Add appropriate amount of previously degassed buffer to reach the desired concentration.
3. Stopper empty glass cuvettes and degas empty (three cycles of 1–2 min of alternating gas and vacuum). Add 2 mL of degassed buffer to the cuvette using a syringe.

5 Enzyme Activity Assays

5.1 POR Assay

1. Catalyzed reaction

$$\text{pyruvate} + \text{CoASH} + \text{MV}_{\text{ox}} \rightarrow \text{acetyl} - \text{CoA} + \text{CO}_2 + \text{MV}_{\text{red}}$$

2. Reagents

Reagents	Amount (per assay)	Final concentration
100 mM EPPS (pH 8.4)	2 mL	100 mM
1 M sodium pyruvate	20 μL	10.0 mM
20 mM coenzyme A	20 μL	0.2 mM
40 mM TPP	20 μL	0.4 mM
200 mM $MgCl_2$	20 μL	2.0 mM
100 mM MV	20 μL	1.0 mM
100 mM DT	Add until blue	Trace

3. Methods

(a) Mix 2 mL of degassed EPPS buffer, 20 μL each of coenzyme A, TPP, $MgCl_2$ and MV (all degassed), and the enzyme sample (~10 μL) to the glass cuvette.

(b) Incubate the mixture at the appropriate temperature in the spectrometer and record at 600 nm ($\varepsilon_{MV} = 13\ \text{mM}^{-1}\ \text{cm}^{-1}$) for ~1 min either digitally or with a chart recorder.

(c) Add 20 μL of degassed pyruvate to start the reaction and continue plotting the absorbance.

(d) Calculate the slopes of the lines before and after the addition of the pyruvate, and then the difference between the two slopes.

(e) Calculate the enzyme activity using the following equation:

$$\text{Activity} = \frac{\Delta\text{OD}}{\text{min}} \times \frac{\text{mM MV} \times \text{cm}}{13.0\ \text{o.d.}} \times \frac{1\ \text{pyruvate oxidized}}{2\ \text{MV reduced}} \times \frac{\text{total volume mL}}{\text{volume of extract mL}}$$

$$U = 1\ \mu\text{mol min}^{-1} \quad \text{specific activity} = \frac{\text{activity}\ (U\ \text{mL}^{-1})}{\text{protein concentration}\ (\text{mg mL}^{-1})} = U\ \text{mg}^{-1}$$

Note: POR activity can also be measured by coupling to the reduction of ferredoxin linked to metronidazole, rather than to reduction of methyl viologen. Add 10 μM ferredoxin and 200 μM metronidazole and follow the decrease at 320 nm. Extinction coefficient of metronidazole = $9.3\ \text{mM}^{-1}\ \text{cm}^{-1}$. Ref. 8.

5.2 H_2ase Activity Assay

1. Catalyzed reaction

$$2MV_{red} + 2H^{+} \rightarrow MV_{ox} + H_2$$

2. Reagents

Reagents	Amount (per assay)	Final concentration
100 mM EPPS (pH 8.4)	2.0 mL	100 mM
1 M DT	20 μL	10 mM
200 mM MV	10 μL	1 mM

3. Methods
   (a) Mix 2.0 mL of degassed EPPS buffer, 10 μL MV and 20 μL DT in a degassed 8 mL serum bottle sealed with a thin septum.
   (b) Incubate the serum bottle at the appropriate temperature for ~2 min prior to adding enzyme sample (5–50 μL). Incubate the mixture at temperature for 2, 4, and 5 min in separate vials.
   (c) Remove the vials and immediately measure the amount of H_2 in the headspace using a gas chromatograph.
   (d) The total amount of H_2 in each vial is calculated using a standard curve generated by adding known amounts of H_2 to vials containing the same buffer without any other reagent. The amounts of H_2 for each time point are plotted against time and the activities are determined from the slope drawn through the points. The activities are normalized by dividing the activity by the protein concentration. One unit (U) of activity is equal to 1 μmol of H_2 produced min^{-1}. Ref. 9.

5.3 Bifurcating H_2ase Activity Assay

1. Catalyzed reaction (four electrons)

$$Fd_{red} + NADH + 3H^{+} \rightarrow Fd_{ox} + NAD^{+} + 2H_2$$

2. Reagents

Reagents	Amount (per assay)	Final concentration
50 mM potassium phosphate (pH 7.5) (anaerobic)	2.0 mL	Same
1 M sodium pyruvate (anaerobic)	20 μL	10 mM
200 mM NADH (anaerobic)	10 μL	1 mM
20 mM thiamine pyrophosphate (TPP)	10 μL	100 μM

(continued)

Reagents	Amount (per assay)	Final concentration
20 mM Flavin mononucleotide (FMN)	10 μL	100 μM
100 mM coenzyme A	20 μL	1 mM
Purified ferredoxin (concentration determined by absorbance at 390 nm; molar absorbance of 17.4 mM^{-1} cm^{-1} per [4Fe4S] cluster)		5 μM
Purified POR	50 μL	

3. Methods
   - (a) Mix 2.00 mL of degassed EPPS buffer, ferredoxin, POR, sodium pyruvate, TPP, FMN, coenzyme A, and NADH in a degassed 8 mL serum bottle sealed with a rubber septum.
   - (b) Incubate the serum bottle at temperature for ~2 min prior to adding sample (5–50 μL). Incubate the mixture at temperature for 2, 4, and 5 min in separate vials.
   - (c) Remove the vials and immediately measure the amount of H_2 in the headspace using a gas chromatograph.
   - (d) The total amount of H_2 in each vial is calculated using a standard curve generated by adding known amounts of H_2 to vials containing the buffer mixture used. The amounts of H_2 for each time point are plotted against time, and the activities are determined from the slope drawn through the points. The activities are normalized by dividing the activity by the protein concentration. One unit (U) of activity is equal to 1 μmol of H_2 produced min^{-1}. Ref. 10.

Notes: This method requires the H_2ase of interest and POR, where POR generates a continuous supply of reduced ferredoxin, which must also be purified. It is very important with this assay to include controls omitting the various components to ensure that a true bifurcating activity is measured. Specifically, the bifurcating H_2ase uses reduced ferredoxin and NADH as the two electron donors, and both are required for hydrogen production. Only insignificant amounts of hydrogen are produced if only one of the electron donors is present

6 Notes

It is absolutely critical to maintain an oxygen-free environment for all of these assays to obtain accurate activity measurements. The most common source of oxygen is a leak in the gas manifold used to degas all components and buffers. For assays involving MV reduction, this can be checked through the addition of trace amounts of DT to the cuvette at the beginning of the assay. If the cuvette is anaerobic, the blue color should remain for a few minutes without fading. If trace amounts of oxygen are present, the solution becomes colorless within seconds. The assay vials for the hydrogenase assay using MV should be dark blue for the duration of the assay.

References

1. Zeldes BM, Keller MW, Loder AJ, Straub CT, Adams MWW, Kelly RM (2015) Extremely thermophilic microorganisms as metabolic engineering platforms for production of fuels and industrial chemicals. Front Microbiol 6:1209
2. Keller MW, Lipscomb GL, Loder AJ, Schut GJ, Kelly RM, Adams MW (2015) A hybrid synthetic pathway for butanol production by a hyperthermophilic microbe. Metab Eng 27:101–106
3. Chandrayan SK, McTernan PM, Hopkins RC, Sun J, Jenney FE Jr, Adams MW (2012) Engineering hyperthermophilic archaeon Pyrococcus furiosus to overproduce its cytoplasmic [NiFe]-hydrogenase. J Biol Chem 287:3257–3264
4. Basen M, Sun J, Adams MW (2012) Engineering a hyperthermophilic archaeon for temperature-dependent product formation. MBio 3:e00053–e00012
5. Copeland WB, Bartley BA, Chandran D, Galdzicki M, Kim KH, Sleight SC, Maranas CD, Sauro HM (2012) Computational tools for metabolic engineering. Metab Eng 14:270–280
6. Thorgersen MP, Lipscomb GL, Schut GJ, Kelly RM, Adams MW (2014) Deletion of acetyl-CoA synthetases I and II increases production of 3-hydroxypropionate by the metabolically-engineered hyperthermophile Pyrococcus furiosus. Metab Eng 22:83–88
7. Lin PP, Mi L, Morioka AH, Yoshino KM, Konishi S, Xu SC, Papanek BA, Riley LA, Guss AM, Liao JC (2015) Consolidated bioprocessing of cellulose to isobutanol using clostridium thermocellum. Metab Engi 31:44–52
8. Blamey JM, Adams MW (1993) Purification and characterization of pyruvate ferredoxin oxidoreductase from the hyperthermophilic archaeon Pyrococcus furiosus. Biochim Biophys Acta 1161:19–27
9. Bryant FO, Adams MW (1989) Characterization of hydrogenase from the hyperthermophilic archaebacterium, Pyrococcus furiosus. J Biol Chem 264:5070–5079
10. Schut GJ, Adams MWW (2009) The iron-hydrogenase of Thermotoga maritima utilizes ferredoxin and NADH synergistically: a new perspective on anaerobic hydrogen production. J Bacteriol 191:4451–4457

Chapter 12

Gene Editing Technologies for Biofuel Production in Thermophilic Microbes

Sharon Smolinski, Emily Freed, and Carrie Eckert

Abstract

Thermophilic microbes are an attractive bioproduction platform due to their inherently lower contamination risk and their ability to perform thermostable enzymatic processes which may be required for biomass processing and other industrial applications. The engineering of microbes for industrial scale processes requires a suite of genetic engineering tools to optimize existing biological systems as well as to design and incorporate new metabolic pathways within strains. Yet, such tools are often lacking and/or inadequate for novel microbes, especially thermophiles. This chapter focuses on genetic tool development and engineering strategies, in addition to challenges, for thermophilic microbes. We provide detailed instructions and techniques for tool development for an anaerobic thermophile, *Caldanaerobacter subterraneus* subsp. *tengcongensis*, including culturing, plasmid construction, transformation, and selection. This establishes a foundation for advanced genetic tool development necessary for the metabolic engineering of this microbe and potentially other thermophilic organisms.

Key words Genetic engineering, Genetic tools, Transformation, Biofuel, Thermophile, *Caldanaerobacter subterraneus* subsp. *tengcongensis*

1 Introduction

A growing number of industrially relevant microbes have been identified and engineered to produce biofuels and value-added chemicals. Of these, thermophiles and extreme thermophiles that can grow at elevated temperatures (above 55 °C and 70 °C, respectively) offer distinct advantages for industrialized processes including lowered risk for contamination, reduced cooling costs, and improved product recovery (especially for volatile products) [1, 2]. In addition, enzymes from thermophilic microbes are active and stable under high temperatures, providing biologically sourced mechanisms for feedstock utilization and catalytic reactions that can also be used in conjunction with nonbiological processes in hybrid systems [1–3].

Michael E. Himmel and Yannick J. Bomble (eds.), *Metabolic Pathway Engineering*, Methods in Molecular Biology, vol. 2096, https://doi.org/10.1007/978-1-0716-0195-2_12,

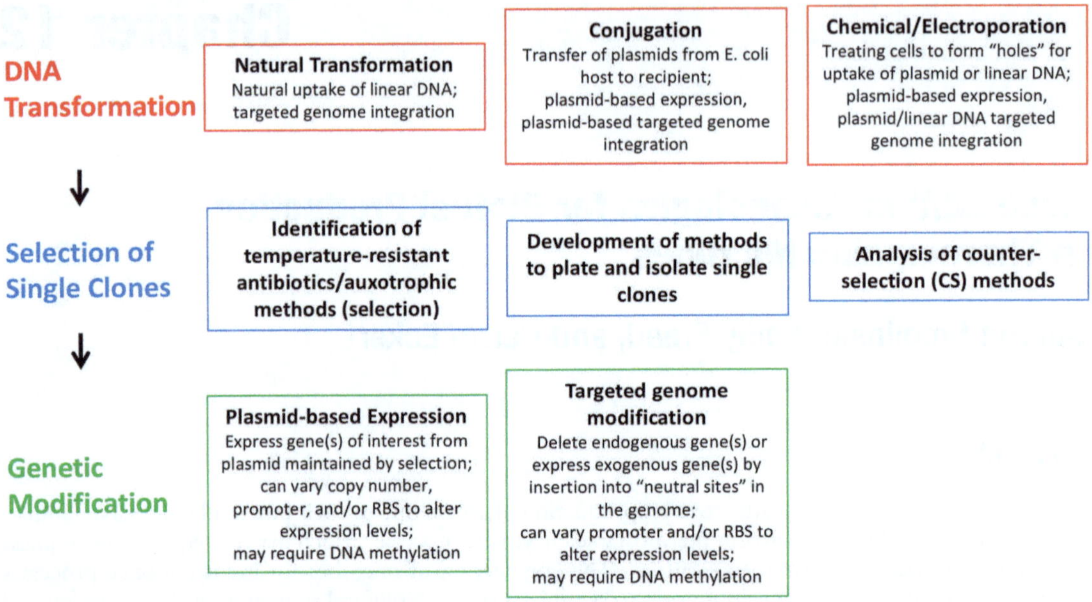

Fig. 1 Overview of pipeline for gene editing

Genetic engineering provides the means to optimize product yields from native metabolic pathways and enhance strain capabilities by the introduction of genes which encode non-native enzymes [1, 2, 4, 5]. For these purposes, genetic tools must be developed to enable targeted and stable genetic alterations. Tool development remains relatively limited for novel, yet industrially valuable organisms, such as thermophiles [2, 4, 5]. For a given strain, researchers must determine suitable methods for introducing genetic material, relevant restriction modification systems, vector development, control of gene and protein expression levels, selection methods, and strategies for plasmid replication or genomic integration (Fig. 1).

Methods for introducing genetic material include natural mechanisms such as natural competence and conjugation, as well as mechanical manipulation using chemically induced competence or electroporation [6]. Phage-enabled transfer may also be possible given the identification and characterization of a growing number of thermophilic phages [7, 8]. If the required cellular machinery is present, natural competence is the most straightforward, only requiring the incubation of a microbe with genetic material under optimized conditions [9–12]. While conjugation is possible in some thermophilic microbes [13, 14], the use of *E. coli* as a carrier is limited due to the higher temperature growth requirements of the strain. While chemical treatment can aid in improving competence [9] and electroporation has been successful for some thermophiles [15–17], the introduced genetic material may be targeted for degradation by any existing restriction and/or modification systems present to protect against foreign DNA. It is therefore helpful to purify DNA for transformation from an *E. coli* strain that matches the methylation pattern of the target microbe. For example, transformation efficiency increases by two orders of magnitude in *Clostridium thermocellum* if the plasmid is purified from *E. coli* that lack Dcm methylation (e.g. BL21 cells) or if Dcm methylation sites in the plasmid are removed [18]. Similarly, expressing methyltransferases from the native organism in *E. coli* can also increase transformation efficiency by at least one order of magnitude, as was demonstrated in *Moorella thermoacetica* [19].

The introduction of genetic material also requires the design of reliable selection strategies. Antibiotics commonly used for selection under mesophilic conditions are often not stable and effective at higher temperatures required for thermophilic growth. In addition, antibiotic resistance genes from mesophilic sources are not always functional at high temperatures, requiring the use of thermostable resistance genes. For example, a thermostable *cat* gene imparting resistance to thiamphenicol [17] and a highly thermostable kanamycin resistance gene (*htk*) [20] that functions up to 75 °C [21] have been successfully employed in thermophilic microbes. Bleomycin [22], hygromycin B [23], and lincomycin [24] have also been used as selection markers in high-temperature systems. If a thermostable antibiotic is not effective for a given strain, selection strategies based on auxotrophy (inability to synthesize a required nutrient) are commonly used [2, 4, 11]. Additionally, selection/counter-selection methods using chemical analogs can also be utilized to select for disruption/loss of genes involved in their toxicity. For example, uracil auxotrophy (loss/mutation of *pyrF* (orotidine-5′-phosphase decarboxylase)) results in tolerance to 5-fluoroorotic acid (5-FOA) [15, 25], while adenine auxotrophy (loss/mutation of adenine phosphotransferase (*apt*)) allows growth in the presence of 6-methyl purine [2, 4]. Argyos et al. demonstrated that loss of a thymidine kinase gene (*tdk*) from *Thermoanaerobacterium saccharolyticum* enabled growth on 5-fluoro-2′-deoxyuridine (FUDR), while disruption of hypoxanthine phosphoribosyl transferase (*hpt*) enabled growth in the presence of purine antimetabolites such as 8-azahypoxanthine (AZH) in *Clostridium thermocellum* [17]. A *lacZ*-counter-selection method has also been described for the thermophile *Bacillus smithii* based on toxicity of high X-gal concentrations in the presence of β-galactosidase activity encoded by a *lacZ*-gene on the integration plasmid [26]. These counterselection methods are useful to select marker-less genome integration events (Fig. 2, see below).

The availability of a sequenced genome is integral to tool development. A bioinformatics approach can mine genomic data for valuable information required for transformation systems, including genes associated with natural competence [27, 28]. Genomic analysis can identify appropriate candidate strains based on

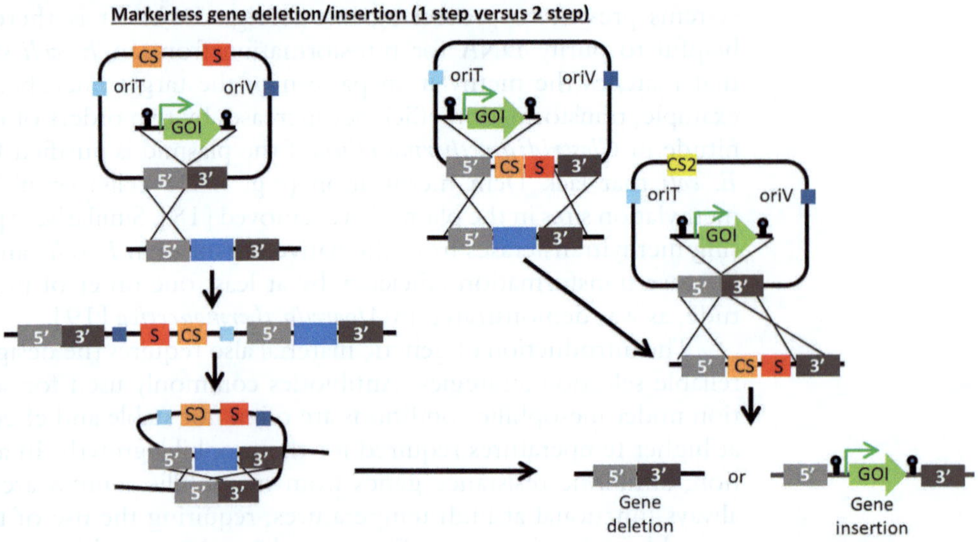

Fig. 2 Gene insertion/deletion by homologous recombination. One-step markerless gene deletion (left) involves one plasmid, containing the targeted homology regions (500–1000 bp 5′ and 3′ of site of insertion/replacement), with or without a GOI cassette (flanked with transcription terminators and driven by a selected promoter), and S and CS on the plasmid backbone with oriT and oriV. Following transformation, a single crossover event occurs, resulting in the integration of the entire plasmid and allowing for selection of integrants via the selectable marker. In a second double crossover recombination event, the plasmid backbone is looped out, thus allowing for selection of these clones using the counter-selection marker. Two-step markerless deletion (right) involves two plasmids with two separate transformations. The first uses 5′ and 3′ homology regions that flank the S and CS ± the GOI cassette. Integration following transformation occurs via double crossover and can be selected for using the selectable marker. The second plasmid contains the 5′ and 3′ homology regions ± the GOI with a second counter selection marker on the backbone. Following transformation into the first isolates, a double crossover event will result in the loss of the first counter selection marker, which can be used for selection of the final desired integration, and loss of the carrier plasmid can be confirmed using the second counter selection marker. *S* selectable marker, *CS* Counter-selectable marker, *GOI* Gene of interest, *oriT* origin of transfer (for conjugation), *oriV* origin of replication

native capabilities by searching for genes encoding native enzymes required for the biosynthesis of a target biofuel or bioproduct and inform strategies for improved fuel production [5]. Gene annotation may not be correct or complete, meaning predicted enzymes and pathways may not function as expected [5]. Bioinformatics also informs vector development and genomic integration, allowing for the identification and use of native replication origins, promoters (ideally of various strengths to facilitate gene expression under various conditions), and sites for genome integration.

Plasmid-based gene expression is commonly used in mesophilic microbes, but there are few stable replicating plasmids available for use in thermophiles [2, 4, 11]. Genome integration allows for targeted gene insertion and deletion in the host strain, allowing for improved stability of genetic modifications and precise engineering [2, 29]. Homologous recombination is the most common approach for genome integration and relies on flanking a marker with sequences that are homologous to regions upstream and downstream of a targeted site on the host genome [2, 29]. The marker can be removed through a double crossover event or counterselection [2, 29] (Fig. 2). This strategy has been used to successfully modify various thermophilic strains, including the disruption of a gene required for histidine biosynthesis [11], the deletion and reintroduction of *pyrF* and insertion of a non-native lactate dehydrogenase gene [15], and the disruption of lactate dehydrogenase and pyruvate formate lyase genes for improved ethanol production [16]. CRISPR-Cas systems provide a new capability for genome integration and editing without the use of selection and counterselection markers [29]. Thermostable Cas9 proteins have been recently identified and utilized for genome editing [30, 31, 32], creating new possibilities for engineering thermophiles.

In this chapter, we present a genetic engineering strategy for *Caldanaerobacter subterraneus* subsp. *tengcongensis*, a thermophilic anaerobe. *C.s.* subsp. *tengcongensis* demonstrates growth between 50 and 80 °C, with optimal growth at 75 °C, and on a variety of substrates (e.g., starch, glucose, maltose, cellobiose, lactose, mannitol, and others) [11, 33]. Genomic analysis indicates carbon monoxide dehydrogenases (CODH) and energy converting hydrogenases (ECH), in addition to a NiFe hydrogenase and a NADH-dependent Fe hydrogenase, as well as multiple glycosidases and esterases [34], some of which have been isolated and characterized as thermostable [35–41]. These features indicate a range of potential biotechnological applications for this strain. Furthermore, this strain is naturally competent, and transformation of a replicating plasmid (using kanamycin resistance), and homologous recombination with plasmid DNA templates have been previously demonstrated [11]. Here, we provide instructions for transforming *C.s.* subsp. *tengcongensis* via natural competence, validating work by Liu et al. [11] and presenting previously unreported details required for successful transformation. We describe materials and methods for culturing, plasmid construction, transformation, and the isolation, verification, and cryopreservation of transformed cells. Not only does this strategy demonstrate the transformability of an anaerobic thermophile, it also serves as a foundation for the development of advanced tools necessary for targeted genetic engineering enabling biofuel production. Specifically, this establishes a base model to develop expanded strategies for plasmid-based expression of genes, targeted gene knock-outs/knock-ins, and the chromosomal incorporation and expression of genes encoding enzymes required for biofuel production (Fig. 3). To this end, our ongoing work includes the assessment of a variety of native promoters for robust and tunable expression, plasmid-based expression of heterologous enzymes, and *pyrF* gene knock-out (via homologous recombination) for subsequent markerless selections (unpublished).

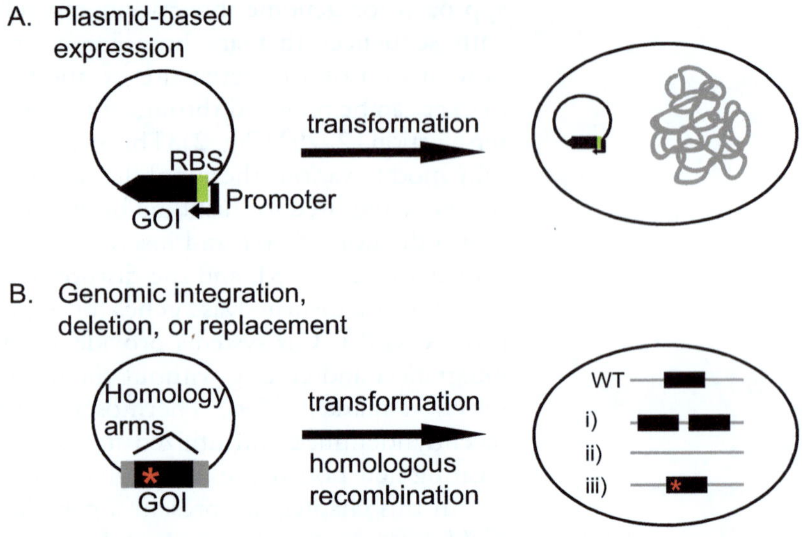

Fig. 3 Strategies for expressing and optimizing the target pathway. (**a**) Plasmid-based expression of a gene(s) of interest (GOI) with a promoter and ribosome binding site (RBS) that allow for constitutive or inducible expression at a specific expression level. The plasmid(s) is maintained by selection. (**b**) Homologous recombination strategies allowing for targeted gene insertion (1) or gene deletion (2) using flanking sequences homologous to the targeted region in the genome. This method can also be used to replace the WT version of a gene, promoter, or RBS with a mutant version if a mutation is known to result in the desired phenotype (3)

2 Materials

2.1 Strains

The wild-type strain *Caldanaerobacter subterraneous* subsp. *tengcongensis* is used for growth and transformation. The strain was obtained from the DSMZ Center (No. 15242). It is a thermophilic anaerobe which was previously classified as *Thermoanaerobacter tengcongensis* [34], before being reclassified as *Caldanaerobacter subterraneus* subspecies *tengcongensis* [42]. For the replication of plasmids, *E. coli* strains should be recombination (*recA*) and endonuclease (*endA*) deficient to allow for improved plasmid stability and quality. Strains JM109 [11] and DH5α were used successfully, and other *recA* and *endA* deficient strains are expected to be appropriate.

2.2 Media

TTE medium (containing starch) was used for growth and transformation [11] (*see* **Note 1**). The composition of the media is provided in Table 1. The media was sterilized by autoclaving (*see* **Note 2**). For solid media, BactoAgar (Sigma-Aldrich) was used (*see* **Note 3**). Filter sterilized sodium thiosulfate was added to the media after autoclaving (*see* **Note 4**). Plates were poured under either anaerobic conditions (in a glove bag) or under aerobic conditions

Table 1
TTE medium (from [11])

	$g\,L^{-1}$
Yeast extract	2.5
NaCl	2.5
Tryptone	5
$(NH_4)_2SO_4$	2
$K_2HPO_4{\cdot}3H_2O$	0.3
$KH_2PO_4{\cdot}7H_2O$	0.3
Cysteine-HCl	0.5
Sodium thiosulfate	5
Starch	10

Adjust pH to 7.5
Add sodium thiosulfate after autoclaving (*see* **Note 4**)
Add 400 mg L^{-1} kanamycin for selection
Potato starch produced the best results, compared with other starch (*see* **Note 2**)

followed by incubation of the solidified plates under anaerobic conditions (in a glove bag). For selection, use 400 mg L^{-1} kanamycin (*see* **Note 5**).

2.3 Gases

Anaerobic gases are required for the growth and transformation of this organism (*see* **Note 6**).

1. 100% N_2.
2. 57% H_2, 35% CO, balanced in argon (*see* **Note 6**).

2.4 Reagents

1. Qiagen DNeasy Blood and Tissue Kit (Qiagen, 69504).
2. NEB Q5 High Fidelity Polymerase Kit (NEB, M0492S).
3. Gibson Assembly Kit (NEB, E2611S).
4. Kanamycin (filter sterilized).

2.5 Equipment

1. Anaerobic chamber (with H_2S scrubber) or glove bag.
2. Anaerobic jar to incubate plates (Mitsubishi™ AnaeroPack™ 2.5 L rectangular jar).
3. Anaerobic gas generating packet (Mitsubishi™ AnaeroPack™ Anaero Anaerobic Gas Generator).
4. Incubator capable of 60 °C.
5. Sparging system to deliver anaerobic gas or gas mixture.
6. Hungate tubes or serum vials.

7. Crimp caps and crimper or stoppers.
8. Sterile syringes and 22–25G needles.
9. Spectrophotometer.
10. Thermocycler.
11. Gel electrophoresis system.
12. Plastic petri plates.

3 Methods

3.1 *Culturing* C.s. *subsp.* Tengcongensis

1. Aliquot 50 mL TTE medium (containing starch) into 160 mL serum bottle(s), add sodium thiosulfate if not already added to medium, and cap the bottle(s) with a gas-tight crimp cap or stopper (*see* **Note 7**).
2. Gas tubes with N_2 (*see* **Note 6**) for 20–30 min (*see* **Note 8**).
3. Inoculate with 0.1 to 1 mL culture or glycerol stock using a sterile needle and syringe.
4. Incubate bottles at 60 °C with shaking (150 rpm).

3.2 *Plasmid Construction*

Plasmid pBluescriptIIKS(+)-cori-tkat is based on pBOL01 [11] and was constructed by inserting the replication origin (OriC) and phosphate acetyltransferase promoter *tte1482* from *C.s.* subsp. *tengcongensis*, and the kanamycin resistance gene *kat*, in pBluescript II KS (+), at the SmaI site of the plasmid backbone using Gibson Assembly (Fig. 4). *C.s.* subsp. *tengcongensis* DNA was prepared using a Qiagen DNeasy Blood and Tissue kit (*see* **Note 9**) for use for PCR amplification of the replication origin and *tte1482* promoter. Primers reported by Liu et al. [11] were used to identify

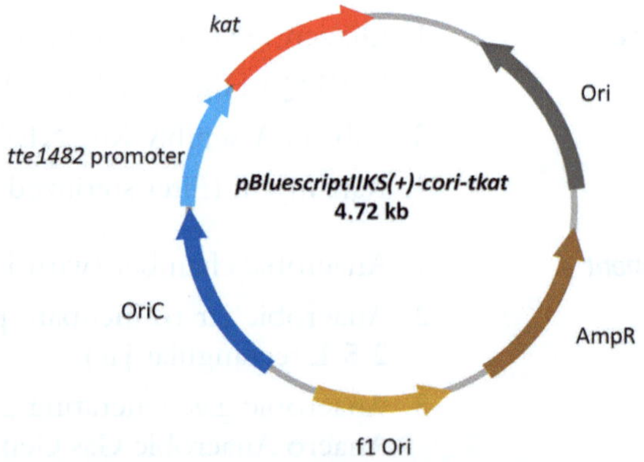

Fig. 4 Plasmid map of pBluescriptIIKS(+)-cori-tkat. The *C.s* subsp. *tengcongensis* replication origin and *tte1482* promoter and *kat* kanamycin resistance gene from pMK18, in pBluescript II KS+. This plasmid is like pBOL01 [11]

Table 2
Primers used for plasmid construction and verification

Procedure	Primer sequence	T_M °C
Amplify fragments for construction of pBluescriptIIKS(+)-cori-tkat by Gibson assembly		
Amplify origin	5′-agaactagtggatccccTAAGCTGTAAGTCTGTGTCTCCC-3′	66.8
	5′-gtgggaatttTCAAAAACAAGGTGTCTTCTGCA-3′	61.2
Amplify *tte1482* promoter	5′-ttgtttttgaAAATTCCCACTCATCACCGCT-3′	60.7
	5′-ggtccattcatTTTAAAATCTCTCCTCCTCTTCGT-3′	60.8
Amplify *kat*	5′-gagattttaaaATGAATGGACCAATAATAATGACT-3′	55.1
	5′-atcgaattcctgcagcccTCAAAATGGTATGCGTTTTGACAC-3′	66.7
Verify presence of kat gene		
Amplify *kat*	5′-ATGAATGGACCAATAATAATGACTAGAG-3′	52.9
	5′-TCAAAATGGTATGCGTTTTGACACA-3′	55.9

Primers for: [1] plasmid construction and [2] verification of the presence of the *kat* gene in transformants. Primers for plasmid construction consist of sequence specific to the fragment being amplified (upper case) and sequence specific to the adjacent fragment (lower case). To verify the presence of the *kat* gene, the primers amplify the full-length gene

the full replication origin sequence in the genome (accessed using Archetype software [43]), and new primers were designed. The phosphate acetyltransferase gene was identified in the genome, and the promoter sequence was verified using BPROM software [44]. Plasmid pMK18 was used as the source of the *kat* gene, which encodes a thermostable adenyl transferase from *Thermus thermophilus* [45], which was provided by Jose Berenguer (Universidad Autonoma de Madrid, Madrid, Spain) (*see* **Note 5**). Plasmid pBluescriptIIKS(+)-cori-tkat was constructed by Gibson Assembly, using fragments amplified with overlapping primers. All primers are shown in Table 2. Ongoing vector development is focused on utilizing promoters of various strengths, *pyrF* disruption (via homologous recombination), and the incorporation and expression of non-native genes for biofuel production (unpublished).

3.3 Transformation Protocol

1. Grow cells to OD 600 nm 0.8–1.2 (*see* **Note 10**).
2. Prepare samples for transformation:
   (a) To 24 mL Hungate tube(s) add 4 mL culture and 4 mL fresh media.
   (b) Add plasmid DNA, 50–100 ng per tube.
   (c) Close tube with stopper or crimp cap.
   (d) Sparge the headspace with N_2 gas (*see* **Note 6**) until anaerobic.
   (e) Incubate 8 h (*see* **Note 11**) at 60 °C (*see* **Note 12**), without shaking (*see* **Note 13**).

3. Transfer samples to solid media.
   (a) Use plates with TTE growth medium containing starch, 2% agar, and 400 mg L^{-1} kanamycin.
   (b) Handle samples under aerobic or anaerobic conditions (i.e., N_2-purged glove bag) (*see* **Note 14**).
   (c) Spread varying amounts of cells on solid media (i.e., 0.1–0.2 mL per petri plate).
   (d) If samples were handled and spread on plates under aerobic conditions, incubate in N_2-purged glove bag for 15–30 min.
   (e) Transfer plates to anaerobic jar.
   (f) Under anaerobic conditions, add anaerobic gas-generating packet and seal jar.
   (g) Incubate at 60 °C in the dark for 60–72 h.

3.4 Isolation and Verification of Transformants

Putative transformants were isolated and regrown on selective media to provide further evidence of successful transformation and provide cells for diagnostic PCR and cryopreservation.

3.4.1 Isolation of Transformants

1. Colonies on selective plates were picked and a portion of the cells were spread on fresh solid media (TTE + 400 mg L^{-1}), for incubation as previously described.
2. Remaining cells were used to inoculate fresh liquid media (TTE + 400 mg L^{-1} kanamycin) in 5 mL serum vials with crimped caps, for sparging and incubation as previously described.

3.4.2 Verification by Diagnostic PCR

1. All steps for diagnostic PCR are done aerobically.
2. Pick a single colony from a selective plate (TTE + 400 mg L^{-1} kanamycin) and use to inoculate 10 μL of ddH_2O in a PCR or 1.5 mL Eppendorf tube (*see* **Note 15**).
3. Boil the inoculated water at 100 °C for 10 min, then place on ice.
4. Set up PCR as per the manufacturer's instructions (*see* **Note 16**). Use 2–4 μL boiled inoculum per 25 μL PCR reaction and primers specific for the plasmid/insert (*see* Table 2). Include DNA from the original plasmid or gene as a positive control.

3.4.3 Cryopreservation

This protocol was developed for the cryopreservation of wild type and genetically modified *C.s.* subsp. *tengcongensis* strains and was successfully used with other anaerobic thermophile strains.

1. Aliquot 0.4 mL 60% glycerol into 2 mL glass vials and seal with a septum.
2. Purge the vials with 100% N_2 until anaerobic.

3. Place the vials in an autoclavable plastic box with a lid to secure the vials (keep upright) and maintain some pressure against the septa (*see* **Note 17**).
4. Autoclave the vials using a 15-min liquid sterilization cycle.
5. Add 0.6 mL dense culture (OD 600 nm > 0.8) per vial using a sterile needle and syringe and mix by gentle shaking.
6. Immediately place at −70 to −80 °C.
7. To use the cryopreserved culture, thaw at room temperature. As soon as the contents are thawed, add entire amount to serum bottle (*see* Subheading 3.1, **step 3**).

4 Notes

1. While *C.s.* subsp. *tengcongensis* is reported to be able to utilize a variety of sugars [33], media containing starch should be used to culture cells for transformation, since the use of other sugars can change cell wall composition and inhibit transformation [11]. Growth was also attempted using medium described in Fardeau et al. [42], but the use of TTE medium resulted in cultures with higher OD.
2. Starch may precipitate as the medium cools after autoclaving; however, potato starch appears to precipitate less than other starch sources.
3. Liu et al. [11] used Gelrite (Sigma-Aldrich), which we did not try. For best results, prepare plates for same-day use.
4. We used a 2 M stock of filter-sterilized sodium thiosulfate, adding 16 mL per 1 L of medium (32 mM final concentration) for growth and transformations. Add to medium immediately before use. Cultures to which sodium thiosulfate is not added exhibit poor growth. For transformations, Liu et al. [11] used a final concentration of 20 mM in the samples of culture diluted in fresh medium.
5. 400 mg L^{-1} kanamycin was used by Liu et al. [11] and was also effective for these studies for the selection of transformants. Our testing additionally showed tolerance to 35 mg L^{-1} thiamphenicol/chloramphenicol, but higher concentrations were not assessed. Previously, studies showed sensitivity to 100 mg L^{-1} of chloromycetin, polymyxin B, streptomycin, and tetracycline and tolerance to 100 mg L^{-1} ampicillin and penicillin [33]. These antibiotics have varying thermostability characteristics, and additional testing can probe the suitability of other antibiotics and concentrations for the sensitivity of *C.s.* subsp. *tengcongensis* to thermostable antibiotics in addition to kanamycin, such as lincomycin and apramycin.

6. For growth and transformation, either 100% N_2 or a mixture of 57% H_2, 35% CO, balanced in argon were used. 100% argon was used for growth, but not tested for transformation. The use of other gases was not tested.
7. Other sizes of serum bottles or Hungate tubes may be used.
8. Establish the time needed to remove O_2 by gassing out samples and measuring samples of the headspace to verify the near-removal of O_2. The time necessary to ensure near-complete removal of O_2 may vary based on flow rate and other parameters specific to the gas delivery system.
9. *C.s.* subsp. *tengcongensis* is classified as Gram-negative [33]. In using the DNeasy Blood and Tissue kit (Qiagen), both the Gram-negative and Gram-positive protocols were used for DNA preparation. Higher yields were obtained from the use of the Gram-negative protocol.
10. This is an experimentally determined effective range of cell density for transformation [11]. Measured OD values can vary between spectrophotometers, and therefore several OD600 values may need to be tested to optimize transformation efficiency.
11. This is an experimentally determined duration of incubation found to be effective for successful transformation, with optimal transformation efficiency identified between 6 and 8 h of incubation [11].
12. The optimal temperatures for transformation have been reported to be 60 °C to 75 °C [11].
13. It is important that cultures are not shaken when incubated with DNA for transformation by natural competence. When cultures are shaken, transformation is not successful, likely due to interference with pili structures from agitation.
14. The handling of samples in aerobic vs. anaerobic conditions during the transfer of culture to liquid and solid media did not yield observed differences, but additional testing and verification may provide further insights. If samples are handled aerobically, they must be made anaerobic by sparging (liquid cultures) or in either a glove box or glove bag (plates) before incubation and growth.
15. Colony PCR to verify transformation was done using colonies picked from solid medium. Attempts at PCR from liquid cultures did not produce amplification products, possibly due to incomplete cell lysis. Method development may be beneficial for resolving issues with the latter approach.
16. Successful PCR reactions have been observed using Q5 High Fidelity DNA Polymerase, Phusion High Fidelity DNA Polymerase, and OneTaq Polymerase using the NEB Tm calculator

(http://tmcalculator.neb.com) to determine annealing temperature for each primer set and polymerase.

17. Best results were achieved by using a few paper towels between the septa and lid and securing the lid in place with tape.

Acknowledgments

This work was supported by the Office of Energy Efficiency and Renewable Energy Bioenergy Technologies Office. This work was authored in part by Alliance for Sustainable Energy, LLC, the Manager and Operator of the National Renewable Energy Laboratory for the U.S. Department of Energy (DOE) under Contract No. DE-AC36-08GO28308. The views expressed in the chapter do not necessarily represent the views of the DOE or the U.S. Government. The U.S. Government retains and the publisher, by accepting the article for publication, acknowledges that the U.S. Government retains a nonexclusive, paid-up, irrevocable, worldwide license to publish or reproduce the published form of this work, or allow others to do so for U.S. Government purposes.

References

1. Turner P, Mamo G, Karlsson EN (2007) Potential and utilization of thermophiles and thermostable enzymes in biorefining. Microb Cell Factories 6:9
2. Zeldes BM, Keller MW, Loder AJ, Straub CT, Adams MWW, Kelly RM (2015) Extremely thermophilic microorganisms as metabolic engineering platforms for production of fuels and industrial chemicals. Front Microbiol 6:1209
3. Yeoman CJ, Han Y, Dodd D, Schroeder CM, Mackie RI, Cann IKO (2010) Thermostable enzymes as biocatalysts in the biofuel industry. Adv Appl Microbiol 70:1–55
4. Lin L, Xu J (2013) Dissecting and engineering metabolic and regulatory networks of thermophilic bacteria for biofuel production. Biotechnol Adv 31:827–837
5. Olson DG, Sparling R, Lynd LR (2015) Ethanol production by engineered thermophiles. Curr Opin Biotechnol 33:130–141
6. Schweitzer HP (2008) Bacterial genetics: past achievements, present state of the field, and future challenges. BioTechniques 44:633–641
7. Liu B, Zhou F, Wu S, Xu Y, Zhang X (2009) Genomic and proteomic characterization of a thermophilic Geobacillus bacteriophage GBSV1. Res Microbiol 160:166–171
8. Nagayoshi Y, Kumagae K, Mori K, Tashiro K, Nakamura A, Fujino Y, Hiromasa Y, Iwamoto T, Kuhara S, Ohshima T, Doi K (2016) Physiological properties and genome structure of the hyperthermophilic filamentous phage φOH3 which infects Thermus thermophilus HB8. Front Microbiol 7:50
9. Koyama Y, Hoshino T, Tomizuka N, Furukawa K (1986) Genetic transformation of the extreme thermophile Thermus thermophilus and of other Thermus spp. J Bacteriol 166:338–340
10. Lipscomb GL, Stirrett K, Schut GJ, Yang F, Jenney FE Jr, Scott RA, Adams MW, Westpheling J (2011) Natural competence in the hyperthermophilic archaeon Pyrococcus furiosus facilitates genetic manipulation: construction of markerless deletions of genes encoding the two cytoplasmic hydrogenases. Appl Environ Microbiol 77:2232–2238
11. Liu B, Wang C, Yang H, Tan H (2012) Establishment of a genetic transformation system and its application in Thermoanaerobacter tengcongensis. J Genet Genomics 39:561–570
12. Shaw AJ, Hogsett DA, Lynd LR (2010) Natural competence in Thermoanaerobacter and Thermoanaerbacterium species. Appl Environ Microbiol 76:4713–4719

13. Cesar CE, Alvarez L, Bricio C, van Heerden E, Littauer D, Berenguer J (2011) Unconventional lateral gene transfer in extreme thermophilic bacteria. Int Microbiol 14:187–199
14. Wahlund TM, Madigan MT (1995) Genetic transfer by conjugation in the thermophilic green sulfur bacterium Chlorobium tepidum. J Bacteriol 177:2583–2588
15. Kita A, Iwasaki Y, Sakai S, Okuto S, Takaoka K, Suzuki T, Yano S, Sawayama S, Tajima T, Kato J, Nishio N, Murakami K, Nakashimada Y (2013) Development of gentic transformation and heterologous expression system in carboxydotrophic thermophilic acetogen Moorella thermoacetica. J Biosci Bioeng 115:347–352
16. Cripps RE, Eley K, Leak DJ, Rudd B, Taylor B, Todd M, Boakes S, Martin S, Atkinson T (2009) Metabolic engineering of Geobacillus thermoglucosidasius for high yield ethanol production. Metab Eng 11:398–408
17. Argyros DA, Tripathi SA, Barrett TF, Rogers SR, Feinberg LF, Olson DG, Foden JM, Miller BB, Lynd LR, Hogsett DA, Caiazza NC (2011) High ethanol titers from cellulose by using metabolically engineered thermophilic, anaerobic microbes. Appl Environ Microbiol 77:8288–8294
18. Guss AM, Olson DG, Caiazza NC, Lynd LR (2012) Dcm methylation is detrimental to plasmid transformation in clostridium thermocellum. Biotechnol Biofuels 5:30
19. Tsukahara K, Kita A, Nakashimada Y, Hoshino T, Murakami K (2014) Genome-guided analysis of transformation efficiency and carbon dioxide assimilation by Moorella thermoacetica Y72. Gene 535:150–155
20. Liao H, McKenzie T, Hageman R (1986) Isolation of a thermostable enzyme variant by cloning and selection in a thermophile. Proc Natl Acad Sci U S A 83:576–580
21. Taylor MP, Esteban CD, Leak DJ (2008) Development of a versatile shuttle vector for gene expression in Geobacillus spp. Plasmid 60:45–52
22. Brouns SJJ, Wu H, Akerboom J, Turnbull AP, de Vos WM, van der Oost J (2005) Engineering a selectable marker for Hyperthermophiles. J Biol Chem 280:11422–11431
23. Atomi H, Imanaka T, Fukui T (2012) Overview of the genetic tools in the Archaea. Front Microbiol 3:337
24. Averhoff B (2006) Genetic systems for Thermus. In: Rainey F, Oren A (eds) Extremophiles (methods in microbiology), vol 35. Oxford Academic Press, Oxford, pp 279–308
25. Tripathi SA, Olson DG, Argyros DA, Miller BB, Barret TF, Murphy DM, McCool JD, Warner AK, Rajgarhia VB, Lynd LR, Hogsett DA, Caiazza NC (2010) Development of pyrF-based genetic system for targeted gene deletion in clostridium thermocellum and creation of a pta mutant. Appl Environ Microbiol 76:6591–6599
26. Bosma EF, van de Weijer AHP, van der Vlist L, de Vos WM, van der Oost J, van Kranenburg R (2015) Establishment of markerless gene deletion tools in thermophilic Bacillus smithii and construction of multiple mutant strains. Microb Cell Factories 20:99
27. Friedrich A, Hartsch T, Averhoff B (2001) Natural transformation in mesophilic and thermophilic bacteria: identification and characterization of novel, closely related competence genes in Acinetobacter sp. strain BD413 and Thermus thermophilus HB27. Appl Environ Microbiol 67:3140–3148
28. Muschiol S, Balaban M, Normark S, Henriques-Normark B (2015) Uptake of extracellular DNA: competence induced pili in natural transformation of *Streptococcus pneumoniae*. BioEssays 37:426–435
29. Liu Z, Liang Y, Ang EL, Zhao H (2017) A new era of genome integration—simply cut and paste! ACS Synth Biol 6:601–609
30. Harrington LB, Paez-Espino D, Staahl BT, Chen JS, Ma E, Kyrpides NC, Doudna JA (2017) A thermostable Cas9 with increased lifetime in human plasma. Nat Commun 8:1424
31. Mougiakos I, Mohanraju P, Bosma EF, Vrouwe V, Finger Bou M, Naduthodi MIS, Gussak A, Brinkman RBL, van Kranenberg R, van der Oost J (2017) Characterizing a thermostable Cas9 for bacterial genome editing and silencing. Nat Commun 8:1647
32. Walker JE, Lanahan AA, Zheng T, Toruno C, Lynd LR, Cameron JC, Olson DG, Eckert CA (2020) Development of both type I–B and type II CRISPR/Cas genome editing systems in the cellulolytic bacterium Clostridium thermocellum. Met Eng Comm 10:e00116
33. Xue Y, Xu Y, Liu Y, Ma Y, Zhou P (2001) Thermoanaerobacter tengcongensis sp. nov., a novel anaerobic, saccharolytic, thermophilic bacterium isolated from a hot spring in Tengcong, China. Int J Syst Evol Microbiol 51:1335–1341
34. Sant'Anna FH, Lebedinsky AV, Sokolova TG, Robb FT, Gonzalez JM (2015) Analysis of three genomes within the thermophilic bacterial species Caldanaerobacter subterraneous with a focus on carbon monoxide

dehydrogenase evolution and hydrolase diversity. BMC Genomics 16:757

35. Abokitse K, Wu M, Bergeron H, Grosse S, Lau PCK (2010) Thermostable feruloyl esterase for the bioproduction of ferulic acid from triticale bran. Appl Microbiol Biotechnol 87:195–203
36. Grosse S, Bergeron H, Imura A, Boyd J, Wang S, Kubota K, Miyadera A, Sulea T, Lau PC (2010) Nature versus nurture in two highly enantioselective esterases from Bacillus cereus and Thermoanaerobacter tengcongensis. Microb Biotechnol 3:65–73
37. Moriyoshi K, Koma D, Yamanaka H, Sakai K, Ohmoto T (2013) Expression and characterization of a thermostable acetylxylan esterase from Caldanaerobacter subterraneous subsp. tengcongensis involved in the degradation of insoluble cellulose acetate. Bioscience. Biotechnol Biochem 77:2495–2498
38. Rao L, Xue Y, Zhou C, Tao J, Li G, Lu JR, May Y (2011) A thermostable esterase from Thermoanaerobacter tengcongesis opening up a new family of bacterial lipolytic enzymes. Biochim Biophys Acta 1814:1695–1702
39. Royter M, Schmidt M, Elend C, Hobenreich H, Schafer T, Bornscheuer UT, Antranikan G (2009) Thermostable lipases from the extreme anaerobic bacteria Thermoanaerobacter thermohydrosulfuricus SOL1 and Caldanaerobacter subterraneous subsp. tengcongensis. Extremophiles 13:769–783
40. Zhang J, Liu J, Zhou J, Ren Y, Dai X, Xiang H (2003) Thermostable esterase from Thermoanaerobacter tengcongensis: high-level expression, purification and characterization. Biotechnol Lett 25:1463–1467
41. Zheng Y, Xue Y, Zhang Y, Zhou C, Schwaneberg U, Ma Y (2010) Cloning, expression, and characterization of a thermostable glucoamylase from Thermoanaerobacter tengcongensis MB4. Appl Microbiol Biotechnol 87:225–233
42. Fardeau ML, Bonilla Salinas M, L'Haridon S, Jeanthon C, Verhe F, Cayol JL, Patel BK, Garcia JL, Olliver B (2004) Isolation from oil reservoirs of novel thermophilic anaerobes phylogenetically related to Thermoanaerobacter subterraneus: reassignment of *T. subterraneus*, Thermoanaerobacter yonseiensis, Thermoanaerobacter tengcongensis and Carboxydibcrachium pacificum to Caldanaerobacter subterraneus gen. nov., sp. nov., comb. nov. as four novel subspecies. Int J Syst Evol Microbiol 54:467–474
43. SGI DNA Archetype [computer software]. SGI DNA, La Jolla, CA
44. Solovyev V, Salamov A (2011) Automatic annotation of microbial genomes and metagenomic sequences. In: Li RW (ed) Metagenomics and its applications in agriculture, biomedicine and environmental studies. Nova Science Publishers, Hauppauge, New York, pp 61–78
45. de Grado M, Castan P, Berenguer J (1999) A high-transformation-efficiency cloning vector for Thermus thermophilus. Plasmid 42:241–245

Chapter 13

Software and Methods for Computational Flux Balance Analysis

Peter C. St. John and Yannick J. Bomble

Abstract

As genetic engineering of organisms has grown easier and more precise, computational modeling of metabolic systems has played an increasingly important role in both guiding experimental interventions and in understanding the results of metabolic perturbations.

Key words Constraint-based reconstruction and analysis, Metabolic networks, Strain design, Flux Balance Analysis, Computational Model

1 Introduction

Flux balance analysis (FBA) is perhaps the most ubiquitous of the computational tools used to study metabolic systems, largely due to its lack of requirement for detailed experimental data and the quality of predictions that it is able to achieve [1]. Also referred as constraints-based reconstruction and analysis (COBRA), FBA predicts the intracellular distribution of metabolic fluxes as a function of a cellular objective and physiological constraints [2]. As opposed to the data-oriented approaches of metabolic flux analysis (*see* Chapter 13), the predictive nature of FBA methods makes them especially suited to solving strain design problems, that is, finding the optimal metabolic intervention to give a desired phenotype.

In a flux balance model, metabolites in a metabolic network are connected through a number of reactions. This system is described via a stoichiometric matrix **S**, such that S_{ij} represents the moles of metabolite i produced (or consumed if $S_{ij} < 0$) by a unit flux through reaction j. A key assumption of the method is that the time scales associated with metabolite interconversion are much faster than those associated with cell growth and substrate uptake,

Michael E. Himmel and Yannick J. Bomble (eds.), *Metabolic Pathway Engineering*, Methods in Molecular Biology, vol. 2096, https://doi.org/10.1007/978-1-0716-0195-2_13,

such that metabolites concentrations are assumed to be in pseudo-steady state. Under this assumption, the fluxes $\mathbf{v}$ through each reaction are governed by the constraints

$$\begin{gathered} \mathbf{S}\mathbf{v} = 0 \\ \mathbf{v}_{\text{lb}} \leq \mathbf{v} \leq \mathbf{v}_{\text{ub}} \end{gathered} \tag{1}$$

where $\mathbf{v}_{lb}$ and $\mathbf{v}_{ub}$ represent the upper and lower bounds of fluxes through each reaction. In this approach, boundaries to the system (exchanges between the cell and the environment) are canonically specified using unbalanced exchange reactions,

$$m_i \leftrightarrow \emptyset \tag{2}$$

Solutions to Eq. (1) span all stochiometrically valid flux profiles, and therefore predictions are further constrained by introducing a cellular objective. This objective is typically the maximization of biomass production, defined using a biomass reaction that consumes precursor metabolites at a rate proportional to their use in biomass synthesis, scaled such that the flux through the reaction is representative of a cell's maximum specific growth rate [3]. By combining an objective function with the stoichiometric constraints in Eq. (1), the fluxes that maximize specific growth rate can be found via linear programming. Methods for solving linear programming problems are highly efficient (almost always completing in polynomial time [4]), even for thousands to tens of thousands of reactions and metabolites. FBA simulations have therefore been able to capture metabolic networks with genome-scale resolution [5] and embedded into design algorithms and bi-level optimization frameworks [6]. Additionally, alternative formulations of the objective function permit different explorations of the flux space to be carried out. Notable examples of these FBA extensions include flux variability analysis (FVA) [7], which characterizes the sensitivity of the optimal solution to individual flux values and minimization of metabolic adjustment (MOMA) [8] and regulatory on/off minimization (ROOM) [9], which attempt to generate more realistic depictions of the flux profile resulting from a genetic perturbation.

Many of the predictions of strain design approaches have been validated experimentally, demonstrating the usefulness of FBA methods for guiding industrial strain development. These techniques typically seek to improve product yields by coupling product production to biomass formation, such that the (new) optimal growth rate is achieved during nonzero product secretion. Knockout strategies from the earliest strain-design method, OptKnock [6], were validated experimentally in the production of lactic acid [10]. Elementary flux modes (EFMs) [11] and minimal cut sets (MCSs) [12] operate based on a geometric interpretation of the FBA flux cone and have led to a number of experimentally successful genetic perturbations [13]. Machado et al. offers a more detailed review of FBA design algorithms [14] and their corresponding experimental implementations.

Software tools for the creation and analysis of COBRA models have substantially improved as the techniques have grown more common. Peer-reviewed genome-scale models are compiled in online databases [15], and online tools exist to aid in converting raw genome sequences into computable metabolic models [16]. Libraries for performing FBA methods have similarly grown in scope and accessibility. Early software libraries include SimPheny™ and CellNetAnalyzer [17] and require commercial or academic licenses. More recent software packages have been released under open source licenses, including the MATLAB-based RAVEN [18] and COBRA toolboxes [19]. This chapter will provide code examples for the COBRApy [20] software package, a library for performing FBA simulations in the open source Python software ecosystem.

This chapter covers three procedures commonly performed during computational flux balance analysis. We first cover the construction of a core carbon metabolic model and the processes involved in validating the model integrity. Next, we discuss several common FBA methods used in predicting flux states, including parsimonious FBA [21] and flux variability analysis while demonstrating how to manipulate model constraints. We next cover the basics of using metabolic models to predict maximum theoretical yields and demonstrate the basics of network visualization.

2 Materials

2.1 Hardware

Computer: No specific hardware requirements are given, and the analyses performed in this chapter should be replicated on any relatively modern laptop or desktop. The installation procedures described below will assume a Unix environment (Mac or Linux), although all steps can be performed on a Windows machine with little to no changes (*see* **Note 1**).

2.2 Software

1. Python environment: The analyses described in this chapter require a Python environment with several packages installed. The Anaconda Python distribution provides a simple method to install Python and several key dependencies on a variety of different operating systems, available from https://www.anaconda.com/download/. This chapter assumes that the user has installed Python 3.6, although a new python environment can be created with the `conda` package manager.
2. Jupyter notebook: Web-based notebooks have become an easy way to interact with python libraries and display results. Using

conda, the jupyter notebook can be installed with `conda install jupyter` and launched using `jupyter notebook`.

3. COBRApy: The main simulation library for performing flux balance calculations in Python. COBRApy can be installed using the Python package manager with `pip install cobra`. Since COBRApy is under active development, some commands may have changed since the writing of this chapter. Using `pip install cobra==0.9.1` will guarantee version compatibility of the commands listed here. We recommend also installing libSBML, which is necessary for parsing older SBML-formatted models with `pip install python-libsbml`.

4. Cameo: A package that builds on COBRApy for performing strain design calculations. Installed via `pip install cameo`.

3 Methods

3.1 Constructing a Core-Carbon Metabolic Model

1. Organize an initial scope of the pathways to be considered. For a core-carbon metabolic model, it is typically critical to consider important reactions in the pentose phosphate pathway, glycolysis, and citric acid cycle. An example python script for specifying a toy metabolic model in Python is listed in Listing 1.

2. For each considered metabolite, assemble the name, chemical formula, charge, and assign a string identifier. The KEGG [22] or MetaCyc [23] databases are particularly useful for finding metabolite and reaction information for previously sequenced genomes or the BIGG database [15] for organisms with a previously developed metabolic model.

3. Combine these metabolites into each reaction needed by the model. For each reaction, determine its corresponding reversibility and set its upper and lower bounds accordingly (*see* **Note 2**). Each reaction is also assigned a corresponding name and string identifier.

4. Verify that each of the core intracellular reactions is mass-balanced (*see* **Note 4**).

5. Add external species and exchange reactions for all species that need to be exchanged with the media. For growth substrates and oxygen, these bounds need to be set to a realistic uptake rate, while other media components (such as nitrogen or carbon dioxide) can often be left unbounded.

6. Construct a reasonable biomass reaction that captures the molecular and energetic requirements for biomass synthesis. For core-carbon models, this reaction typically ensures that all necessarily central-carbon intermediates (i.e., pyruvate, acetyl-

CoA, citric acid cycle, and pentose-phosphate products) can be synthesized in appropriate quantities. Stoichiometry for these intermediates can be found by summing over the biomass composition of each key amino acid and accounting for synthesis pathways (*see*, for instance, McKinlay et al. [24]). For models that include these pathways explicitly, amino acid stoichiometry can be included explicitly.

7. Add an ATP maintenance reaction to capture the nongrowth-associated energy cost of cell repair. This reaction is typically specified with a nonzero, positive lower bound to force flux through the reaction.
8. Ensure the model can produce biomass successfully with reasonable bounds on import and export reactions. This step typically involves adding missing cofactor exchange reactions, reactions associated with electron transport, and nonenzyme catalyzed equilibrium reactions (*see* **Note 5**).
9. Tailor the bounds on important reactions such that FBA predictions match ^{13}C flux data and biomass growth rates as closely as possible.

Listing 1: Example Python Code for Creating and Optimizing a Simple Metabolic Model:

```
import cobra

glc = cobra.Metabolite(id='glc', name='D-Glucose',
                       formula='C6H12O6', charge=None)
pyr = cobra.Metabolite(
    id='pyr', name='Pyruvate', formula='C3H3O3')
pep = cobra.Metabolite(
    id='pep', name='Phosphoenolpyruvate', formula='C3H2O6P')
g6p = cobra.Metabolite(
    id='g6p', name='D-Glucose 6-phosphate', formula='C6H11O9P')

# Construct a COBRA Reaction
GLCpts = cobra.Reaction(
    id='GLCpts', name='D-glucose transport via PEP:Pyr PTS',
    lower_bound=0., upper_bound=1000.)

# Specify the reaction stoichiometry
GLCpts.add_metabolites(
    {glc: -1.0, pep: -1.0, g6p: 1.0, pyr: 1.0})

# Ensure the reaction is mass-balanced
assert GLCpts.check_mass_balance() == {}

# Add a number of other reactions to complete the model.
# *Note*: These reactions are not mass-balanced
upper_glycol = cobra.Reaction(id='upg')
lower_glycol = cobra.Reaction(id='lwg')
atp_maint    = cobra.Reaction(id='atpm')
pyr_kin      = cobra.Reaction(id='pyk')
```

```
atp = cobra.Metabolite('atp', formula='C10H16N5O13P3')
adp = cobra.Metabolite('adp', formula='C10H15N5O10P2')
co2 = cobra.Metabolite('co2', formula='CO2')

upper_glycol.add_metabolites({g6p: -1, pyr: 1, pep: 1, atp: 1, adp: -1})
lower_glycol.add_metabolites({pyr: -1, adp: -3, co2: 3, atp: 3})
atp_maint.add_metabolites({atp: -1, h2o: -1, adp: 1})
pyr_kin.add_metabolites({pyr: -1, adp: -1, atp: 1, pep: 1})

# Create the COBRA model object and add each reaction.
# Note, metabolites are added automatically.
test_model = cobra.Model('test')
test_model.add_reactions([upper_glycol, lower_glycol, atp_maint,
GLCpts])

# Add exchanges for the boundary species
co2_bound = test_model.add_boundary(test_model.metabolites.co2)
glc_bound = test_model.add_boundary(test_model.metabolites.glc)
glc_bound.lower_bound = -10

# Specify a model objective and optimize the model using FBA
test_model.objective = test_model.reactions.atpm
test_model.optimize()
```

3.2 Simulating the Phenotype Phase Plane for E. coli *Growth Under Variable Oxygen Flux*

1. Load or construct a metabolic model of *E. coli* central carbon metabolism. In this example, we use `EColiCore` [25]. Code to reproduce this analysis in COBRApy is given in Listing 2.
2. Set realistic uptake and exchange rates for all boundary reactions except oxygen.
3. While allowing oxygen uptake to take any value, perform FBA to find the maximum oxygen uptake rate at full aerobic growth. Using this value as an upper bound, divide oxygen flux into a number of intervals between the maximum O_2 flux and zero.
4. At each interval, set the maximum oxygen uptake flux to the designated bound.
5. Perform parsimonious flux balance analysis (pFBA) [21] to find the most likely flux distribution. In pFBA, the model is solved twice. In the first optimization, the maximum objective value is stored. In the second optimization, the objective value is constrained to its maximum value, while the absolute values of the remaining fluxes are minimized (*see* **Note 6**). These fluxes are stored for each value of oxygen flux.

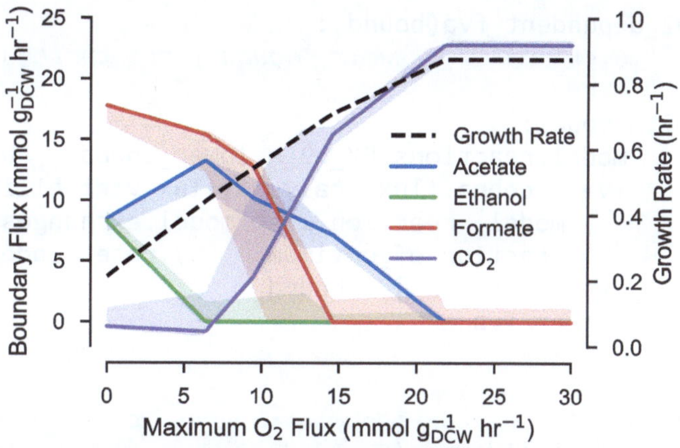

Fig. 1 Flux variability analysis for the phenotype phase plane of *E. coli* under varying oxygen conditions. Solid lines indicate the minimum-norm solutions as calculated with pFBA, while shaded regions indicate accessible flux states with a 0.95 FVA fraction. Growth rate is plotted on a separate *y*-axis on the right

6. Perform flux variability analysis (FVA) for each oxygen flux. In FVA, the objective value of the model is constrained to a fractional value of its maximum (*see* **Note** 7). Flux through each reaction is then iteratively maximized and minimized to determine its feasible range.
7. The results of this analysis can then be plotted against the one (or more) varied fluxes. Results for oxygen-dependent flux of exchange reaction in `EColiCore` are shown in Fig. 1.

Listing 2: Code Sample to Calculate the Data Needed the Phenotype Phase Plane Visualization Shown in Fig. 1:

```
import numpy as np
import pandas as pd

# Load the E Coli Core test model
import cobra.test
model = cobra.test.create_test_model("textbook")

def o2_dependent_flux(bound):
    """ Given an O2 flux in `bound`, return the corresponding pFBA
solution
    """

    # Use a cobrapy context manager to handle the reversible
manipulation of
    # model bounds
    with model:
        model.reactions.EX_o2_e.lower_bound = -bound
        return cobra.flux_analysis.pfba(model).fluxes
```

```
def o2_dependent_fva(bound):
    """ Given an O2 flux in `bound`, return the corresponding FVA bounds
    """
    with model:
        model.reactions.EX_o2_e.lower_bound = -bound
        fva = cobra.flux_analysis.flux_variability_analysis(
            model, reaction_list=model.exchanges,
            fraction_of_optimum=0.99).reset_index()
        fva['o2_bound'] = bound
        return fva


# Define a list of oxygen flux bounds to iterate over
o2_bound = np.linspace(0, 30, 200)

# Save the list of pFBA and FVA solutions to a pandas Dataframe
o2_fluxes = pd.concat([o2_dependent_flux(bound) for bound in o2_bound],
                      axis=1, ignore_index=True).T
o2_fva = pd.concat([o2_dependent_fva(bound) for bound in o2_bound],
axis=0)
```

3.3 Calculating Maximum Theoretical Yield and Optimal Metabolic Pathway for Cis-Cis-Muconate Synthesis

1. Load or construct a metabolic model of *P. putida* central carbon metabolism. In this example, we use a previously constructed metabolic model. Code to reproduce this analysis in COBRApy is given in Listing 3.
2. For maximum yield calculations, ATP maintenance costs are typically neglected. Bounds that force ATP consumption are therefore set to zero.
3. The objective of the FBA model is set to the export reaction for the desired metabolite (in this case, the muconate sink reaction). The model is subsequently optimized.
4. Molar yields are calculated by dividing the molar flux of inputs (glucose) by that of the output (muconate). Carbon-mole and mass-weighted yields are calculated by scaling this ratio by the respective carbon content and molecular weights.
5. Visualization of the resulting flux distribution can be obtained via `d3flux`. An example output of this tool is shown in Fig. 2 (*see* **Note 8**).

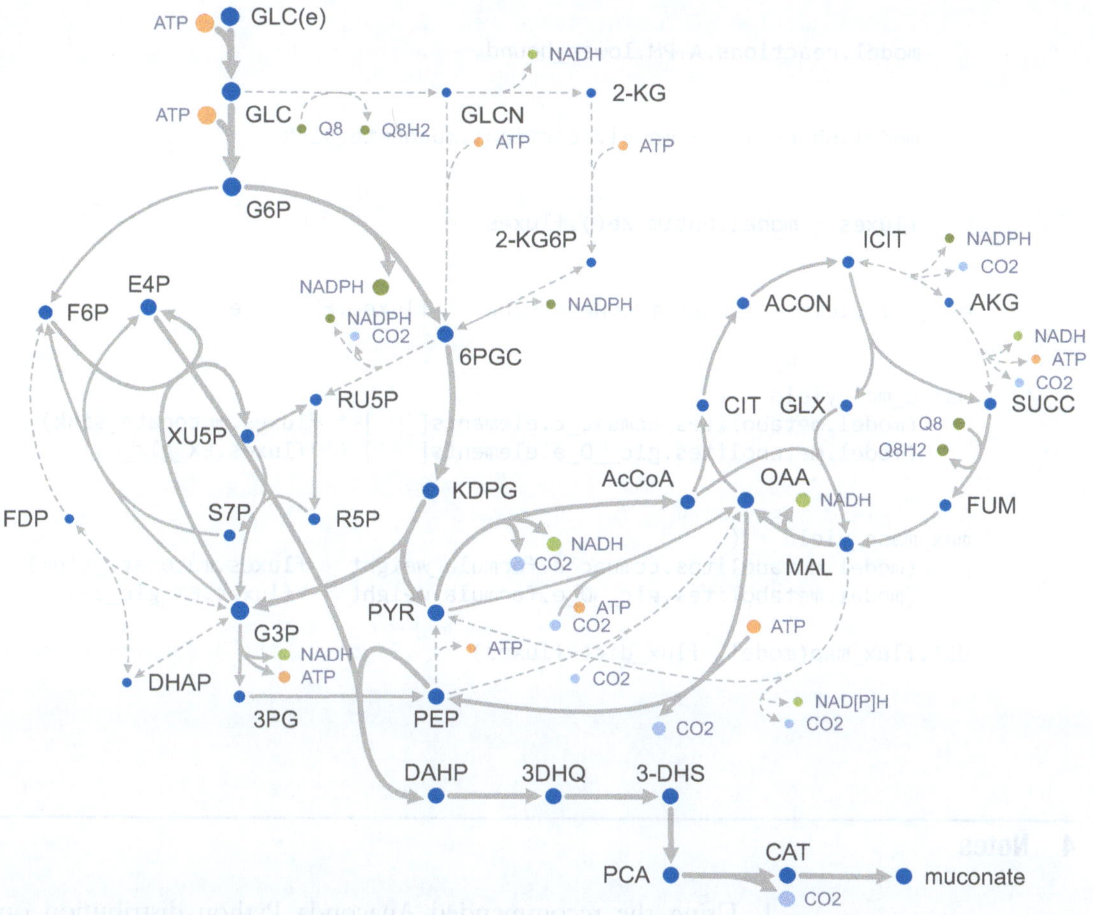

Fig. 2 Visualization of the pathway with the maximum muconate yield, as computed via Listing 3. This pathway was visualized using d3flux

Listing 3: Code to Calculate the Maximum Theoretical Yield of Muconate from Glucose for a Core-Carbon Model of *P. putida*: Note, the code used to import the model has been replaced by a placeholder

```
import cobra
import d3flux as d3f

model = cobra.io.load_json_model('...')

with model:
```

```
    # Zero out the ATP maintenance reaction
    model.reactions.ATPM.lower_bound = 0.

    # Set the objective reaction to the product of interest
    model.objective = model.reactions.muconate_sink

    # Optimize and store maximum yield flux vector
    fluxes = model.optimize().fluxes

# Calculate maximum molar yield
max_mol_yield = fluxes.muconate_sink / -fluxes.EX_glc_e

# Calculate carbon yield
max_c_mol_yield = (
    (model.metabolites.ccmuac_c.elements['C'] * fluxes.muconate_sink) /
    (model.metabolites.glc__D_e.elements['C'] * -fluxes.EX_glc_e))

# Calculate mass yield
max_mass_yield = (
    (model.metabolites.ccmuac_c.formula_weight * fluxes.muconate_sink) /
    (model.metabolites.glc__D_e.formula_weight * -fluxes.EX_glc_e))

d3f.flux_map(model, flux_dict=fluxes)  # Visualize the resulting pathway
```

4 Notes

1. Using the recommended Anaconda Python distribution on windows provides an "Anaconda Prompt" utility accessible from the start menu. This utility allows the user to enter package installation commands and start the jupyter notebook server in a manner similar to the instructions given for Unix terminals. Notable exceptions are commands that begin with `source activate ...` for mac and linux terminals should be replaced with `activate ...` on windows machines.
2. For unbounded reactions, an upper or lower bound of ± 1000 is typically used.
3. For reactions that are not immediately mass balanced, the most common solution is typically to balance missing hydrogen or charges with water.
4. Because of the difficulty associated with developing a working model from scratch, it is often substantially easier to begin with a working model for a similar organism, and simply adapt the enzymatic reactions and biomass stoichiometry to match the

currently studied organism. However, investigating a failed optimization can often go a long way toward indicating the particular reaction that is missing (or mis-specified). Tools such as the shadow prices or reduced costs are particularly useful, as well as investigating blocked metabolites and reactions on a one-by-one basis.

5. In optimization problems where absolute values are minimized in the objective function, the problem can typically remain linear through the introduction of slack variables. Reusing the original solution basis can therefore make the two successive solutions pFBA relatively efficient.
6. A fraction of 0.95 is often used, although 0.99 can also be used if the ranges generated with an FVA fraction of 0.95 are much wider than those observed experimentally. Fixing the FVA fraction to 1.0 is useful in diagnosing model identifiability or equivalent pathways.
7. D3Flux is available from github.com/pstjohn/d3flux. A more fully featured metabolic visualization library is available in `escher` [26]. However, metabolic network visualization is inherently a time-consuming and highly specific task requiring manual positioning of nodes and annotations in most cases.

Acknowledgments

This work was funded by the U.S. Department of Energy's Bioenergy Technologies Office (DOE-BETO). This work was authored in part by Alliance for Sustainable Energy, LLC, the Manager and Operator of the National Renewable Energy Laboratory for the U.S. Department of Energy (DOE) under Contract No. DE-AC36-08GO28308. The views expressed in the chapter do not necessarily represent the views of the DOE or the U.S. Government. The U.S. Government retains, and the publisher, by accepting the article for publication, acknowledges that the U.S. Government retains a nonexclusive, paid-up, irrevocable, worldwide license to publish or reproduce the published form of this work, or allow others to do so, for U.S. Government purposes.

References

1. Orth JD, Thiele I, Palsson BØ (2010) What is flux balance analysis? Nat Biotechnol 28:245–248. https://doi.org/10.1038/nbt.1614
2. Lewis NE, Nagarajan H, Palsson BO (2012) Constraining the metabolic genotypephenotype relationship using a phylogeny of in silico

methods. Nat Rev Microbiol. https://doi.org/10.1038/nrmicro2737

3. Feist AM, Palsson BO (2010) The biomass objective function. Curr Opin Microbiol 13:344–349. https://doi.org/10.1016/j.mib.2010.03.003
4. Spielman DA, Teng S-H (2004) Smoothed analysis of algorithms. J ACM 51:385–463. https://doi.org/10.1145/990308.990310
5. Thiele I, Palsson BØ (2010) A protocol for generating a high-quality genome-scale metabolic reconstruction. Nat Protoc 5:93–121. https://doi.org/10.1038/nprot.2009.203
6. Burgard AP, Pharkya P, Maranas CD (2003) Optknock: a bilevel programming framework for identifying gene knockout strategies for microbial strain optimization. Biotechnol Bioeng 84:647–657. https://doi.org/10.1002/bit.10803
7. Mahadevan R, Schilling C (2003) The effects of alternate optimal solutions in constraint-based genome-scale metabolic models. Metab Eng 5:264–276. https://doi.org/10.1016/j.ymben.2003.09.002
8. Segre D, Vitkup D, Church GM (2002) Analysis of optimality in natural and perturbed metabolic networks. Proc Natl Acad Sci U S A 99:15112–15117. https://doi.org/10.1073/pnas.232349399
9. Shlomi T, Berkman O, Ruppin E (2005) Regulatory on/off minimization of metabolic flux changes after genetic perturbations. Proc Natl Acad Sci U S A 102:7695–7700. https://doi.org/10.1073/pnas.0406346102
10. Fong SS, Burgard AP, Herring CD et al (2005) In silico design and adaptive evolution of *Escherichia coli* for production of lactic acid. Biotechnol Bioeng 91:643–648. https://doi.org/10.1002/bit.20542
11. Zanghellini J, Ruckerbauer DE, Hanscho M, Jungreuthmayer C (2013) Elementary flux modes in a nutshell: properties, calculation and applications. Biotechnol J 8:1009–1016. https://doi.org/10.1002/biot.201200269
12. Hädicke O, Klamt S (2011) Computing complex metabolic intervention strategies using constrained minimal cut sets. Metab Eng 13:204–213. https://doi.org/10.1016/j.ymben.2010.12.004
13. Shen CR, Lan EI, Dekishima Y et al (2011) Driving forces enable high-titer anaerobic 1-butanol synthesis in *Escherichia coli*. Appl Environ Microbiol 77:2905–2915. https://doi.org/10.1128/aem.03034-10
14. Machado D, Herrgård MJ (2015) Co-evolution of strain design methods based on flux balance and elementary mode analysis. Metab Eng Commun 2:85–92. https://doi.org/10.1016/j.meteno.2015.04.001
15. King ZA, Lu J, Dräger A et al (2015) BiGG models: a platform for integrating, standardizing and sharing genome-scale models. Nucleic Acids Res 44:D515–D522. https://doi.org/10.1093/nar/gkv1049
16. Henry CS, DeJongh M, Best AA et al (2010) High-throughput generation, optimization and analysis of genome-scale metabolic models. Nat Biotechnol 28:977–982. https://doi.org/10.1038/nbt.1672
17. Klamt S, Saez-Rodriguez J, Gilles ED (2007) BMC Syst Biol 1:2. https://doi.org/10.1186/1752-0509-1-2
18. Agren R, Liu L, Shoaie S et al (2013) The RAVEN toolbox and its use for generating a genome-scale metabolic model for *penicillium chrysogenum*. PLoS Comput Biol 9:e1002980. https://doi.org/10.1371/journal.pcbi.1002980
19. Hyduke D, Hyduke D, Schellenberger J et al (2011) COBRA toolbox 2.0. Protocol Exchange. https://doi.org/10.1038/protex.2011.234
20. Ebrahim A, Lerman JA, Palsson BO, Hyduke DR (2013) COBRApy: COnstraints-based reconstruction and analysis for python. BMC Syst Biol 7:74. https://doi.org/10.1186/1752-0509-7-74
21. Lewis NE, Hixson KK, Conrad TM et al (2010) Omic data from evolved *E. coli* are consistent with computed optimal growth from genome-scale models. Mol Syst Biol 6. https://doi.org/10.1038/msb.2010.47
22. Kanehisa M, Furumichi M, Tanabe M et al (2016) KEGG: new perspectives on genomes, pathways, diseases and drugs. Nucleic Acids Res 45:D353–D361. https://doi.org/10.1093/nar/gkw1092
23. Caspi R, Altman T, Billington R et al (2013) The MetaCyc database of metabolic pathways and enzymes and the BioCyc collection of pathway/genome databases. Nucleic Acids Res 42:D459–D471. https://doi.org/10.1093/nar/gkt1103
24. McKinlay JB, Shachar-Hill Y, Zeikus JG, Vieille C (2007) Determining actinobacillus succinogenes metabolic pathways and fluxes by NMR and GC-MS analyses of ^{13}C-labeled metabolic product isotopomers. Metab Eng 9:177–192.

https://doi.org/10.1016/j.ymben.2006.10.006

25. Orth JD, Palsson BØ, Fleming RMT (2010) Reconstruction and use of microbial metabolic networks: the core *Escherichia coli* metabolic model as an educational guide. EcoSal Plus 4. https://doi.org/10.1128/ecosalplus.10.2.1

26. King ZA, Dräger A, Ebrahim A et al (2015) Escher: a web application for building, sharing, and embedding data-rich visualizations of biological pathways. PLoS Comput Biol 11: e1004321. https://doi.org/10.1371/journal.pcbi.1004321

Chapter 14

Dynamic Flux Analysis: An Experimental Approach of Fluxomics

Wei Xiong, Huaiguang Jiang, and PinChing Maness

Abstract

Metabolic flux analysis represents an essential perspective to understand cellular physiology and offers quantitative information to guide pathway engineering. A valuable approach for experimental elucidation of metabolic flux is dynamic flux analysis, which estimates the relative or absolute flow rates through a series of metabolic intermediates in a given pathway. It is based on kinetic isotope labeling experiments, liquid chromatography-mass spectrometry (LC-MS), and computational analysis that relate kinetic isotope trajectories of metabolites to pathway activity. Herein, we illustrate the mathematic principles underlying the dynamic flux analysis and mainly focus on describing the experimental procedures for data generation. This protocol is exemplified using cyanobacterial metabolism as an example, for which reliable labeling data for central carbon metabolites can be acquired quantitatively. This protocol is applicable to other microbial systems as well and can be readily adapted to address different metabolic processes.

Key words Cell harvesting, Dynamic flux analysis, Experimental metabolomics, Isotope tracer, Quenching, LC-MS, Metabolic flux

1 Introduction

The emergence of the tools in synthetic biology and knowledge in genomic sciences has opened opportunities for designing chassis organisms to produce renewable fuels and chemicals. Yet, realizing this potential is not straightforward largely due to the complexity of interacting metabolic networks, the underlying regulations and signaling, as well as pathway bottleneck(s). Another technology gap is the redox and energy imbalance often encountered in microbes undergoing major genomic modifications leading to unstable mutants. ^{13}C-based fluxomics described herein could address these technology gaps and provide a breakthrough solution to design, modify, and optimize microbes (or plants and biomes) for beneficial purposes. With a tight relevance to guide metabolic engineering, fluxomics affords an integrated experimental/computational approach that is designed to systematically quantify the

Michael E. Himmel and Yannick J. Bomble (eds.), *Metabolic Pathway Engineering*, Methods in Molecular Biology, vol. 2096, https://doi.org/10.1007/978-1-0716-0195-2_14,

rates of biochemical reactions within a biological entity [1]. In its earliest appearance, fluxomics was a computational methodology based on Flux Balance Analysis (FBA) [2], in which intracellular fluxes are merely estimated from stoichiometric reaction models with a handful of experimental inputs (e.g., extracellular consumption and secretion rates). More elaborate fluxomic analysis emerged afterward, which employs isotope tracer experiments and quantitative isotope analysis by means of Mass Spectrometry (MS) and/or Nuclear Magnetic Resonance (NMR) measurements [3]. A typical example is Dynamic Flux Analysis (DFA), also termed kinetic flux profiling [4]. This approach relies on the dynamics of cellular incorporation of stable isotope from substrate into downstream metabolites. The half-time labeling of a metabolite will then depend directly on the speed of transmission of label into the metabolite (i.e., the flux) and inversely on the size of the metabolite pool (i.e., the intracellular metabolite concentration) [4]. By analyzing the labeling half-time ($t_{0.5}$) of a metabolite and $t_{0.5}$ of its precursors/products, the intracellular process within a metabolic pathway can be assessed accordingly. In contrast to steady-state labeling strategy, DFA provides metabolite labeling trajectories as a function of time. It has the special ability to identify bottlenecks in a cascading pathway, a premise for genome-scale engineering and targeted biodesign aimed at improving flux for microbial synthesis of advanced biofuels. In addition, it is particularly suitable for addressing photoautotrophic metabolism assimilating carbon solely from CO_2 which would otherwise produce a uniformed steady-state ^{13}C-labeling pattern that is not sensitive to fluxes [5]. In recent years, DFA approach has been used to investigate a number of important metabolic processes. It addressed nitrogen assimilation fluxes in *E. coli* and a non-model photosynthetic microbe [6, 7], unveiled the topological structure of the tricarboxylic acid cycle in cyanobacteria [8, 9], and identified the rate-limiting reactions in an engineered isoprene biosynthetic pathway [10] and others.

Experiments for dynamic metabolite labeling can work independently for qualitative understanding of pathway activities. Moreover, they can also serve as experimental inputs for ordinary differential equation (ODE) models, allowing quantitative estimation of intracellular carbon fluxes [11]. The basic concept of this algorithm is demonstrated in Fig. 1 and described in details in Subheading 1.1. Although detailed computational methodology may go beyond the scope of this chapter, it is noteworthy that progress has been made in developing computational tools. Publicly available software (e.g., INCA [12]) has enabled experimental biologists with basic computational skills to calculate intracellular fluxes from kinetic labeling data. Please refer to Yuan et al. [4] for computational protocols of DFA. In this chapter, we mainly focus on the key experimental steps necessary for DFA, and we use

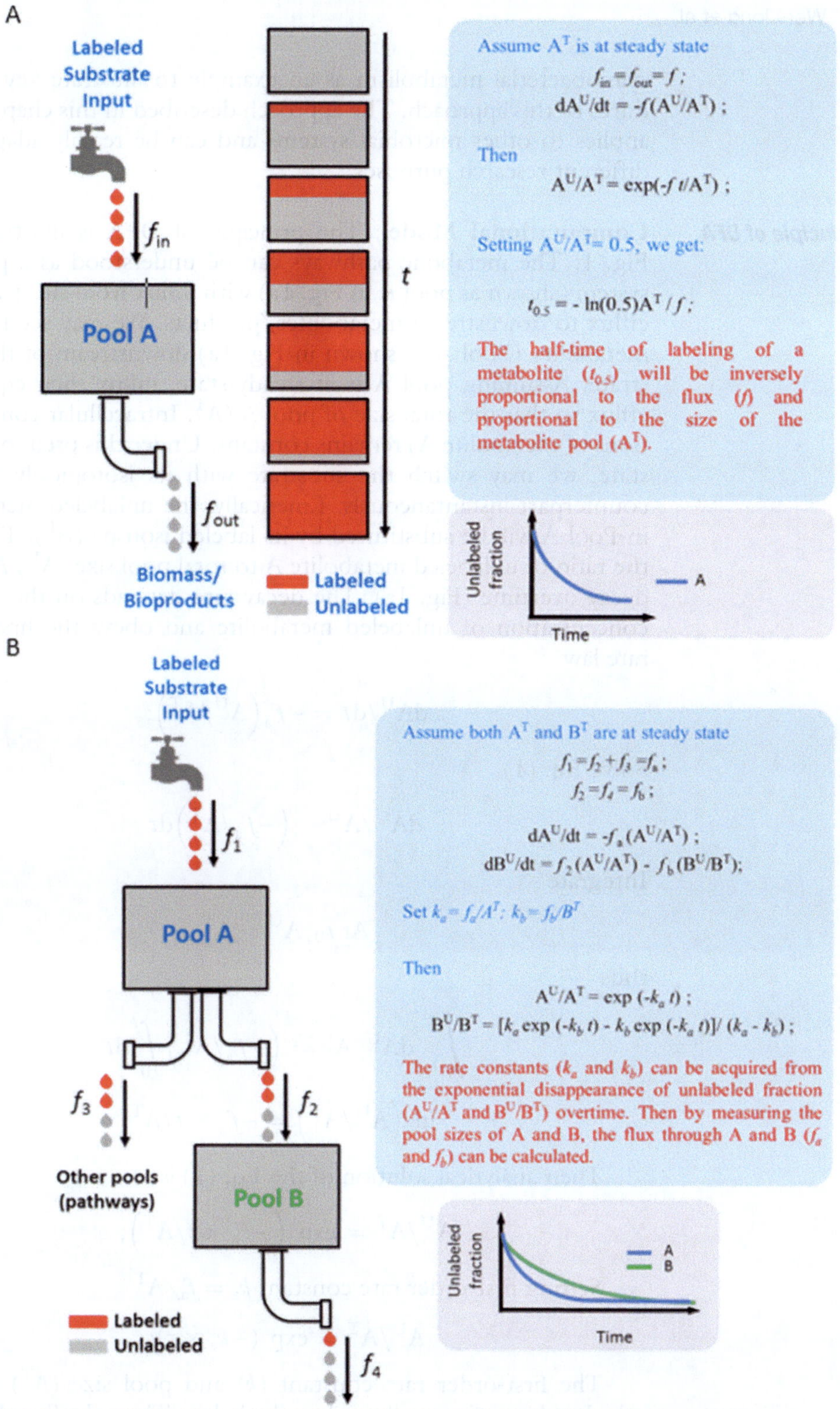

Fig. 1 Computational principle of dynamic flux analysis. (**a**) A simplified scenario: metabolite A is synthesized directly from substrate by f_{in} and it is also used for synthesizing biomass or other bioproducts with constant flux rate ($f_{in} = f_{out}$); (**b**) Imagine a more complicated case in which metabolite B is located downstream of A, and the flux through metabolite B can be obtained accordingly. The mathematical deductions are illustrated in blue boxes and Subheading 1.1

cyanobacterial metabolism as an example to illustrate key procedures of this approach. The approach described in this chapter also applies to other microbial systems and can be readily adapted to different research purposes.

1.1 Principle of DFA

Computational Model: The principle of DFA is illustrated in Fig. 1. The metabolic pathways can be understood as a pooling system (shown as pool A in Fig. 1a) with influx from substrate and efflux to downstream metabolites/products. We may see the first metabolite (Pool A as shown in Fig. 1a) downstream of the substrate. Assuming pool A is at steady-state, influx then equals to efflux so that the total size of pool A (A^T, Intracellular concentrations of metabolite A) remains constant. Under this pseudo-steady state, we may switch the substrate with its isotopically labeled counterpart instantaneously. Kinetically, the unlabeled metabolite in Pool A will be substituted by its labeled isotope (A^L). Thereby, the ratio of unlabeled metabolite A to total pool size (A^U/A^T) will decay overtime (Fig. 1a). The decay rate depends on the relative concentration of unlabeled metabolite and obeys the first-order rate law.

$$dA^U/dt = -f_a\left(A^U/A^T\right); \tag{1}$$

Solve Eq. (1).

$$dA^U/A^U = \left(-f_a/A^T\right)dt$$

Integrate

$$\text{At } t_0, A^U = A^T$$

thus

$$\int_{A^T}^{A^U} dA^U/A^U = \left(-f_a/A^T\right)\int_0^t dt$$

$$\ln\left(A^U/A^T\right) = -f_a \times t/A^T$$

Then analytical solution of the Eq. (1) is:

$$A^U/A^T = \exp\left(-f_a \times t/A^T\right); \tag{2}$$

Setting first-order rate constant $k_a = f_a/A^T$

$$A^U/A^T = \exp\left(-k_a \times t\right); \tag{3}$$

The first-order rate constant (k) and pool size (A^T) can be calculated experimentally as describe below. Thus, the flux through metabolite A (f_a) can be quantified (Fig. 1a).

Next, let's consider the flux through downstream metabolite. Imagining another pool (Pool B, Fig. 1b) located downstream of

Pool A and a proportion of f_a directs to B (f_2). The decay of unlabeled metabolite B (dB^U/dt) will be delayed by labeling of A, in that the influx (f_2) obtains a proportion of unlabeled mass from A. This unlabeled proportion decays exponentially overtime and correspondingly:

$$dB^U/dt = f_2\left(A^U/A^T\right) - f_b(B^U/B^T); \quad (4)$$

Setting first-order rate constant

$$k_a = f_a/A^T, k_b = f_b/B^T;$$

Then analytical solution is:

$$B^U/B^T = [k_a \exp(-k_b \times t) - k_b \exp(-k_a \times t)]/(k_a - k_b); \quad (5)$$

Measuring k_a and k_b experimentally as well as A^T and B^T, flux through A and B (f_a and f_b) will be obtained, respectively (*see* **Note 1**).

Experimental measurements: Based on computational model described above, two key parameters are required to be determined experimentally: the intracellular concentration of the target metabolites (e.g., A^T and B^T) and their rate constants (e.g., k_a and k_b). Both can be measured by metabolomics approaches through stable isotope-tracer and mass spectrometry analysis. The procedures for these measurements are illustrated in Fig. 2. Briefly, for measuring absolute concentration of intracellular metabolites, we adopt a reverse-labeling strategy by which cellular metabolites are completely labeled through ^{13}C-substrate feeding in advance and subsequently spiked with unlabeled standards during extraction and LC-MS analysis. The absolute concentration of metabolites can be obtained by quantitatively comparing the intensities of

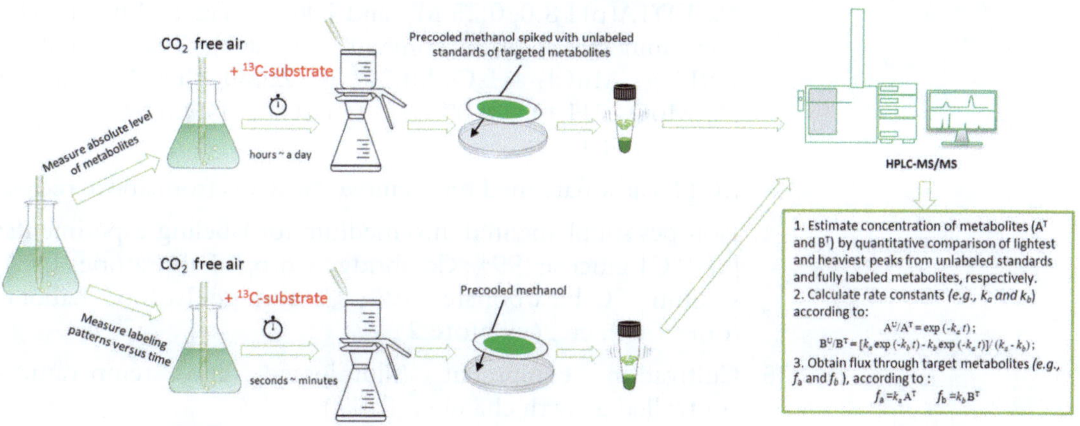

Fig. 2 Procedures for Dynamic Flux Analysis (DFA) exemplified by cyanobacterial cells fed with ^{13}C-substrate

labeled isotopes which are from intracellular metabolites and unlabeled isotopes that are derived from standards.

Measurement of rate constants can be realized by feeding cells with isotope-labeled substrate, fast sampling, and quenching of cells, followed by metabolite extraction and LC-MS analysis. For reliable measurements, labeling status at given time points should be captured accurately. It requires prompt cell harvesting and rapid quenching of their metabolism, otherwise the metabolic state of those fast turnover pathways will not be truly recorded. For this purpose, we developed a vacuum filtration approach for cell accumulation. It allows cell to be harvested within a few seconds. The filters with attached cells can be immersed into precooled organic solvent (e.g., methanol). This step is effective in ceasing immediate metabolism due to low temperature and inactivation of metabolic enzymes by denaturing. Afterward, metabolites will be extracted and analyzed by LC-MS. The fraction of unlabeled metabolite in total pool could be quantitatively analyzed by comparing the light and heavier isotope ratios of a metabolite in MS. The rate constants obtained accordingly can be combined with intracellular concentrations of metabolites to estimate fluxes.

2 Materials

1. Strain: *Synechocystis* PCC 6803.
2. BG11 medium for *Synechocystis* growth (per liter of final medium): 10 mL100× BG-11, 1 mL 6 mg/mL Ferric ammonium citrate, 1 mL 20 mg/mL Na_2CO_3, 1 mL 30.5 mg/mL K_2HPO_4. The 100× BG-11 stock solution contains (per liter): 149.6 g $NaNO_3$, 7.5 g $MgSO_4 \cdot 7H_2O$, 3.6 g $CaCl_2 \cdot 2H_2O$, 0.60 g Citric acid (or 0.89 g Na-citrate, dihydrate), 1.12 mL Na-EDTA(pH 8.0, 0.25 M), and 100 mL Trace Minerals. The trace minerals solution contains (per liter): 2.86 g H_3BO_3, 1.81 g $MnCl_2 \cdot 4H_2O$, 0.22 g $ZnSO_4 \cdot 7H_2O$, 0.39 g $Na_2MoO_4 \cdot 2H_2O$, 0.079 g $CuSO_4 \cdot 5H_2O$,0.049 g $Co(NO_3)_2 \cdot 6H_2O$.
3. BG11-carbonate medium: remove Na_2CO_3 from above recipe.
4. Isotopes supplemented into medium for labeling experiments: [U-^{13}C]-glucose (99%; Cambridge Isotope Laboratories Inc.); sodium ^{13}C-bicarbonate (99%; Cambridge Isotope Laboratories Inc.), etc. (*see* **Note 2**).
5. Cultivation equipment (illuminated and temperature-controlled growth chamber, flasks).
6. Photometer (e.g., LI-COR Model LI-250 light meter) used for measuring light intensity in *Synechocystis* growth chamber.

7. Spectrometer (e.g., WPA Biowave II spectrometer) for measuring optical density (OD) of *Synechocystis* at 730 nm in plastic cuvettes.
8. 47-mm nylon membrane filters for harvesting cells (Whatman Nylon Membrane 0.8 μm) and vacuum filtration apparatus (e.g., Nalgene 300–4050 polysulfone graduated filter holder with receiver, 47 mm diameter, 500 mL capacity).
9. Tweezers for manipulating the membrane filters.
10. MS-grade water.
11. MS-grade methanol.
12. Dry ice.
13. −80 °C freezer.
14. Microcentrifuge for pelleting cell debris during extraction.
15. Nitrogen evaporation system (e.g., Glas-Col ZipVap Zanntek Analytical Evaporator).
16. Standards for any metabolites of interest (e.g., glucose-6-phosphate).
17. HPLC-ESI-MS/MS (e.g., Dionex UltiMate 3000 HPLC linked to Bruker MicrOTOF-Q) (*see* **Note 3**).
18. HPLC column: Synergi Hydro-RP (C18) 250 mm × 1 mm column (Phenomenex, Aschaffenburg, Germany).
19. Ion pairing reagent: tributylamine (e.g., Sigma-Aldrich, cat. no. T49352).
20. Consumables (Eppendorf tubes, HPLC vials and caps, etc.)

3 Methods

3.1 Quantify Absolute Concentrations of Targeted Metabolites

3.1.1 Cell Cultivation, Isotope Labeling, Sampling, and Quenching

1. *Synechocystis* PCC 6803 precultures are bubbled with CO_2-free air at 30 °C in 500 mL BG11-carbonate medium supplemented with a carbon source (e.g., 5 mM glucose) under 50 μE s^{-1} m^{-2} white light illuminations (*see* **Note 4**).
2. When cells are at mid-exponential growth phase with an OD_{730nm} between 0.8 and1.0, dilute the culture by 1:20 in a new 500-mL flask with 500 mL fresh BG11-carbonate. Bubble the culture with CO_2-free air by flowing the air through a 5 M fresh sodium hydroxide solution first to remove CO_2.
3. For ^{13}C-labeling, supplement the *Synechocystis* culture with ^{13}C-substrate (e.g., U-^{13}C-glucose, 99% purity at a final concentration of 5 mM).
4. Wait until cells grow up when OD_{730nm} reaches between 0.8 and 1.0. Aspirate 50 mL culture and pass it through the membrane filter by vacuum. The cyanobacterial cells attached

membrane is immediately and completely immersed into 1 mL dry-ice precooled methanol in a petri-dish.

5. Place the dish in −80 °C freezer for 30–60 min to quench cell metabolism.

3.1.2 Metabolite Extraction

1. Remove samples from −80 °C freezer and place them on an icebox filled with dry ice. Add mixture of metabolite standards into the methanol at various given concentrations.
2. On each petri-dish, scrape the cells off the membrane with a cell scraper. Transfer the cell suspension into a 2-mL Eppendorf tube. Centrifuge for 1 min at 10,000 × *g* at 4 °C to pellet the cells. Transfer the supernatant into a new 1.5-mL EP tube and set aside at 4 °C.
3. Resuspend the pellet in 300 μL of extraction solution (50:50 of methanol: water) by vortexing and keep it at 4 °C for 5 min. Centrifuge for 1 min at 10,000 × *g* and 4 °C, combine the supernatant with the supernatant obtained from **step 1** (*see* **Note 5**).
4. Repeat **step 3** for a final round of extraction.
5. Transfer the cell extract into HPLC vials.
6. Evaporate the extract solutions under N_2 gas flow until dry and resuspend in 200 μL extract solution (*see* **Note 6**).

3.1.3 Metabolite Analysis by HPLC-MS/MS

1. HPLC setup: A Synergi Hydro-RP (C18) 250 mm × 1 mm column (Phenomenex, Aschaffenburg, Germany) is utilized for chromatographic separation. The mobile phase was comprised of eluent A (10 mM tributylamine in water with 3% methanol, adjusted pH to 4.95 with 15 mM acetic acid) and eluent B (methanol) with the following linear gradients: $t = 0$, 0% B; $t = 5$ min, 0% B; $t = 20$ min, 20% B; $t = 35$ min, 20% B; $t = 60$ min, 65% B; $t = 65$ min, 95% B; $t = 75$ min, 95% B; $t = 80$ min, 0% B. The injection volume was 5 μL and before each injection, the column was equilibrated by pre-running 100% eluent A for 3 min. The flow rate was 0.05 mL/min (*see* **Note 7**).
2. MS setup: Metabolites can be detected by different types of MS detectors. Here we use Bruker MicrOTOF-Q mass spectrometer operated in the negative-ion mode as an example. Relevant instrument settings for the MicrOTOF-Q are as follows: capillary, +4000 V; end plate offset, −500 V; funnels 1 and 2 RFs, 200 *V*pp; hexapole RF, 140 *V*pp; quadrupole ion energy, 5 eV; low mass, 55 *m/z*; collision energy, 10 eV; collision RF, 100 *V*pp; transfer time, 75 μs; prepulse storage, 4 μs. Instrument settings for the Ion Trap were: capillary, +3500 V; end plate offset, −500 V; nebulizer, 2.8 bar; dry gas flow, 8.0 L min^{-1}; dry temperature, 350 °C; low mass, 55 *m/z*.

The MS data were processed using Bruker Data Analysis 4.0 software.

3. Prepare mixture of metabolite standards in HPLC vials and load the vials into the LC-MS autosampler. Apply the samples of standards to HPLC-MS/MS with settings described in **steps 1–2** and determine the LC-MS parameters for each metabolite (e.g., retention time, mass-to-charge ratio (*m/z*)) (*see* Table 1).

Table 1
Specific metabolite-dependent parameters in LC-MS

	Metabolites	Formula	M.W.	(M–H) ^{12}C	(M–H) ^{13}C	RT (min)
Sugar phosphate	Glucose	$C_6H_{12}O_6$	180.06	179.06	185.08	3.1
	Glucose-6-phosphate	$C_6H_{13}O_9P$	260.03	259.02	265.04	17.7
	Fructose-6-phosphate	$C_6H_{13}O_9P$	260.03	259.02	265.04	18.6
	Fructose 1,6-bisphosphate	$C_6H_{14}O_{12}P_2$	340	338.99	345.01	45.2
	Dihydroxyacetone phosphate	$C_3H_7O_6P$	170	168.99	172	49.9
	Glyceraldehyde 3-phosphate	$C_3H_7O_6P$	170	168.99	172	19.9
	3-phosphoglycerate	$C_3H_7O_7P$	185.99	184.99	188	48.6
	Diphosphoglycerate	$C_3H_8O_{10}P_2$	265.96	264.95	267.96	55.9
	Phosphoenolpyruvate	$C_3H_5O_6P$	167.98	166.98	169.99	50.6
	6-phosphogluconate	$C_6H_{13}O_{10}P$	276.02	275.02	281.04	47.4
	Ribulose-5-phosphate	$C_5H_{11}O_8P$	230.02	229.01	234.03	16.3
	Ribulose-1,5-biphosphate	$C_5H_{12}O_{11}P_2$	309.99	308.98	314	50.5
	Xylulose-5-phosphate	$C_5H_{11}O_8P$	230.02	229.01	234.03	19.9
	Ribose-5-phosphate	$C_5H_{11}O_8P$	230.02	229.01	234.03	18.2
	Sedoheptulose-7-phosphate	$C_7H_{15}O_{10}P$	290.04	289.03	296.05	18.3
	Sedoheptulose-1,7-biphosphate	$C_7H_{16}O_{13}P_2$	370.01	369	376.02	48.2
	Erythrose 4-phosphate	$C_4H_9O_7P$	200.01	199	203.01	21.5
	2-C-Methyl-D-erythritol 2,4-cyclodiphosphate	$C_5H_{12}O_9P_2$	278	276.98	282	50.4
	(2*E*)-4-hydroxy-3-methylbut-2-*en*-1-yl trihydrogen diphosphate	$C_5H_{12}O_8P_2$	262	260.99	266.01	51.8
Organic acids	Citric acid	$C_6H_8O_7$	192.03	191.02	197.04	50.3
	Isocitric acid	$C_6H_8O_7$	192.03	191.02	197.04	50.3
	Alpha-ketoglutaric acid	$C_5H_6O_5$	146.02	145.01	150.03	46.5
	Succinic acid	$C_4H_6O_4$	118.03	117.02	121.03	34.7
	Fumaric acid	$C_4H_4O_4$	116.01	115	119.01	48.5
	Oxaloacetic acid	$C_4H_4O_5$	132.01	131	135.01	53.5
	Malic acid	$C_4H_6O_5$	134.02	133.01	137.02	42.1
	Glyoxylic acid	$C_2H_2O_3$	74	72.99	75	15.9
	Pyruvic acid	$C_3H_4O_3$	88.02	87.01	90.02	23
	Acetoacetic acid	$C_4H_6O_3$	102.03	101.02	105.03	30.5
	Succinic acid semialdehyde	$C_4H_6O_3$	102.03	101.02	105.03	12.3

(continued)

Table 1
(continued)

	Metabolites	Formula	M.W.	(M–H) ^{12}C	(M–H) ^{13}C	RT (min)
Amino acids	L-Arginine	$C_6H_{14}N_4O_2$	174.11	173.1	179.12	2.3
	L-Asparagine	$C_4H_8N_2O_3$	132.05	131.05	135.06	3
	L-Aspartic acid	$C_4H_7NO_4$	133.04	132.03	136.04	9.8
	L-Cysteine	$C_3H_7NO_2S$	121.02	120.01	123.02	3.2
	L-Glutamic acid	$C_5H_9NO_4$	147.05	146.05	151.07	7.1
	L-Glutamine	$C_5H_{10}N_2O_3$	146.07	145.06	150.08	3.1
	Glycine	$C_2H_5NO_2$	75.03	74.02	76.03	3.1
	L-Histidine	$C_6H_9N_3O_2$	155.07	154.06	160.08	2.4
	L-Isoleucine	$C_6H_{13}NO_2$	131.09	130.09	136.11	5.8
	L-Leucine	$C_6H_{13}NO_2$	131.09	130.09	136.11	6.3
	L-Lysine	$C_6H_{14}N_2O_2$	146.11	145.1	151.12	2.3
	L-Methionine	$C_5H_{11}NO_2S$	149.05	148.04	153.06	4.7
	L-Phenylalanine	$C_9H_{11}NO_2$	165.08	164.07	173.1	13.9
	L-Proline	$C_5H_9NO_2$	115.06	114.07	119.09	3.3
	L-Serine	$C_3H_7NO_3$	105.04	104.03	107.04	3
	L-Threonine	$C_4H_9NO_3$	119.06	118.05	122.06	3.1
	L-Tyrosine	$C_9H_{11}NO_3$	181.07	180.07	189.1	6.7
	L-Valine	$C_5H_{11}NO_2$	117.08	116.07	121.09	4
	L-Trypthopan	$C_{11}H_{12}N_2O_2$	204.09	203.08	204.08	20.6
	Trans-4-hydroxy-L-proline	$C_5H_9NO_3$	131.06	130.05	135.07	3.1
	Gama-aminobutyric acid	$C_4H_9NO_2$	103.06	102.06	106.07	2.5
	L-citrulline	$C_6H_{14}N_3O_3$	175.1	174.09	180.11	3.1
	L-ornithine	$C_5H_{13}N_2O_2$	132.09	131.08	136.1	3.1
	L-Pyroglutamate	$C_5H_7NO_3$	129.04	128.04	133.06	18.2
	5(*d*)-Aminolevulinic Acid	$C_5H_9NO_3$	131.06	130.05	135.07	2.6
	N-Acetyl-L-glutamate	$C_7H_{12}NO_5$	189.06	188.06	195.08	42.1
	N-Acetyl-L-glucosamine	$C_8H_{15}NO_6$	221.09	220.08	228.11	3.1
Redox cofactors	FMN	$C_{17}H_{21}N_4O_9P$	456.1	455.1	472.16	53.4
	FAD	$C_{27}H_{33}N_9O_{15}P_2$	785.16	784.15	811.24	56
	NAD	$C_{21}H_{27}N_7O_{14}P_2$	663.11	662.1	683.17	24.5
	NADH	$C_{21}H_{29}N_7O_{14}P_2$	665.12	664.12	685.19	51.7
	NADP	$C_{21}H_{29}N_7O_{17}P_3$	743.08	742.07	763.14	50
	NADPH	$C_{21}H_{31}N_7O_{17}P_3$	745.09	744.08	765.15	56.3
Coenzyme A	CoA	$C_{21}H_{36}N_7O_{16}P_3S$	767.12	766.11	787.18	
	Acetyl CoA	$C_{23}H_{38}N_7O_{17}P_3S$	809.13	808.12	831.2	58.7
	Malonyl CoA	$C_{24}H_{38}N_7O_{19}P_3S$	853.12	852.11	876.19	59
	Acetoacetyl CoA	$C_{25}H_{40}N_7O_{18}P_3S$	850.13	851.13	876.21	58.5
	Succinyl CoA	$C_{25}H_{40}N_7O_{19}P_3S$	867.13	866.12	891.2	58.9
	Butyryl CoA	$C_{25}H_{42}N_7O_{17}P_3S$	837.16	836.16	861.24	61.6

(continued)

Table 1
(continued)

	Metabolites	Formula	M.W.	(M–H) ^{12}C	(M–H) ^{13}C	RT (min)
Nucleotides related	Cyclic adenosine diphosphate ribose	$C_{15}H_{21}N_5O_{13}P_2$	541.06	540.05	555.1	25
	dGMP	$C_{10}H_{14}N_5O_7P$	347.06	346.06	356.09	33.3
	dATP	$C_{10}H_{16}N_5O_{12}P_3$	491	489.99	500.02	56.2
	dTTP	$C_{10}H_{17}N_2O_{14}P_3$	481.99	480.98	491.01	55.5
	dGTP	$C_{10}H_{16}N_5O_{13}P_3$	507	505.99	516.02	55.1
	dCTP	$C_9H_{16}N_3O_{13}P_3$	466.99	465.98	475.01	54.6
	Adenosine	$C_{10}H_{13}N_5O_4$	267.1	266.09	276.12	17.9
	Thymidine	$C_{10}H_{14}N_2O_5$	242.09	241.08	251.12	15.8
	Guanosine	$C_{10}H_{13}N_5O_5$	283.09	282.08	292.12	13.9
	Cytidine	$C_9H_{13}N_3O_5$	243.09	242.08	251.11	3.3
	Uridine	$C_9H_{12}N_2O_6$	244.07	243.06	252.09	7.6
	ATP	$C_{10}H_{16}N_5O_{13}P_3$	507	505.99	516.02	55.8
	ADP	$C_{10}H_{14}N_5O_{10}P_2$	427.03	426.01	436.04	53.2
	AMP	$C_{10}H_{15}N_5O_7P$	347.06	346.04	356.07	33.5
	GTP	$C_{10}H_{16}N_5O_{14}P_3$	522.99	521.98	532.02	55.1
	GDP	$C_{10}H_{15}N_5O_{11}P_2$	443.02	442.02	452.05	49.3
	GMP	$C_{10}H_{14}N_5O_8P$	363.06	362.05	372.08	27
	CTP	$C_9H_{16}N_3O_{14}P_3$	482.98	481.98	491.01	54.3
	CDP	$C_9H_{15}N_3O_{11}P_2$	403.02	402.01	411.04	46.5
	CMP	$C_9H_{14}N_3O_8P$	323.05	322.04	331.07	23.3
	UTP	$C_9H_{15}N_2O_{15}P_3$	483.97	482.96	491.99	55.1
	UDP	$C_9H_{14}N_2O_{12}P_2$	404	402.99	412.03	49.1
	UMP	$C_9H_{13}N_2O_9P$	324.04	323.03	332.06	25.5
	UDP-D-glucose	$C_{15}H_{24}N_2O_{17}P_2$	566.06	565.05	580.1	47.1
	UDP-GlcNAC	$C_{17}H_{27}N_3O_{17}P_2$	607.08	606.07	623.13	47.2
	UDP-D-xylose	$C_{14}H_{22}N_2O_{16}P_2$	536.04	535.04	549.08	47.3
	UDP-*N*-acetylmuraminate	$C_{20}H_{31}N_3O_{19}P_2$	679.1	678.1	698.16	54.6
Others	Cyanopterin	$C_{20}H_{28}N_5O_{13}$	547.18	546.17	566.24	19
	Glucosyl glycerol	$C_9H_{18}O_8$	254.1	253.09	262.12	3.3
	Glutathione (reduced)	$C_{10}H_{17}N_3O_6S$	307.08	306.08	316.11	20.1
	Glutathione (oxidized)	$C_{20}H_{32}N_6O_{12}S_2$	612.15	611.14	631.21	42.3

4. Analyze the samples (prepared in Subheading 3.1.2, **step 6** or Subheading 3.2.2, **step 6**) by LC-MS, using 5-μL injection volume (increase the volume to 25 μL if necessary; overload may result in increased ion suppression and poor LC separation performance), preprogrammed LC gradient and MS settings.

3.1.4 Calculate the Intracellular Concentration of Metabolites

1. According to the MS data of each metabolite, calculate the ratio of its peak intensity to the intensity of corresponding unlabeled standard. $R = \text{Int.}_{met}/\text{Int.}_{std}$. An example is illustrated in Fig. 3.

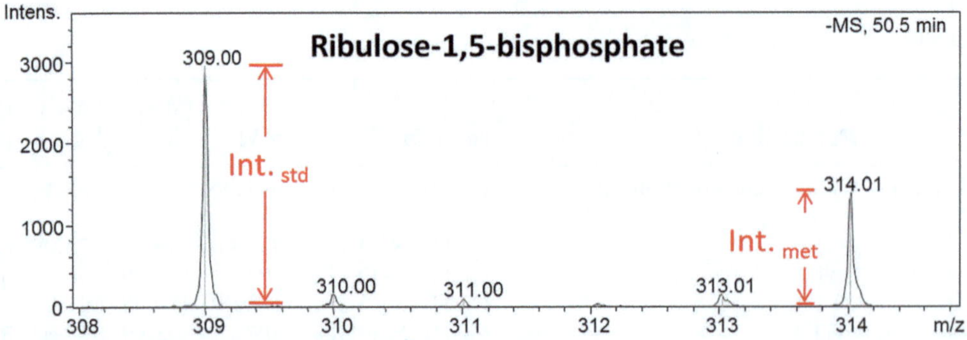

Fig. 3 Schematic of MS peak intensities obtained via Subheadings 3.1.2 and 3.1.3 for ^{13}C-labeled intracellular metabolite (met) with its unlabeled standard (std). Exemplified by Ribulose-1,5-bisphosphate (RuBP), peak at *m*/*z* of 309.00 represents unlabeled internal standard of RuBP in negative mode and peak at 314.01 represents fully labeled RuBP (M+5) derived from cells after long-term ^{13}C labeling. Since the concentration of the internal standard is known, the intracellular concentration of RuBP can be estimated by the ratio of peak intensities. ($R = \text{Int.}_{\text{met}}/\text{Int.}_{\text{std}}$) Herein, $\text{Int.}_{\text{met}} = 1391$, $\text{Int.}_{\text{std}} = 3092$; therefore $R = 0.45$

2. Calculate the concentration of metabolite in a cell by:

$$C = (R \times A)/\left(V_{\text{sampling}} \times \text{OD} \times n \times V_{\text{cell}}\right)$$

C: Concentration of a metabolite in a single cell;

A: Absolute amount of a standard in a sample;

V_{sampling}: Volume of cell culture taken for a sample (e.g., Here we take 50 mL culture as described in Subheading 3.1.1, **step 4**);

OD: Optical density of the culture;

n: Cell numbers per OD. For *Synechocystis*, 1-mL culture contains 10^8 cells per $\text{OD}_{730\text{ nm}}$;

V_{cell}: Average volume of a cell. The average volume of a *Synechocystis* cell is 3 fL or 3×10^{-15} L (*see* **Note 8**).

3.2 Measure Disappearance of Unlabeled Fraction of Metabolites Overtime

3.2.1 Cell Cultivation, Kinetic Isotope Labeling, Sampling, and Quenching

1. *Synechocystis* PCC 6803 precultures are bubbled with CO_2-free air at 30 °C in 500 mL BG11-carbonate medium supplemented with carbon source (e.g., 5 mM glucose) under 50 μE s^{-1} m^{-2} white light illuminations (*see* **Note 4**).
2. Cells are harvested at mid-exponential growth phase at an $\text{OD}_{730\text{nm}}$ between 0.8 and 1.0 by centrifugation at 4000 × *g* and room temperature for 5 min. Discard the supernatant and resuspend cell pellet on 500 mL fresh BG11-carbonate medium by adding 100 mM bicarbonate. Equilibrate the culture in growth chamber for 10 min.
3. Aspirate 50 mL culture and pass it through the membrane filter by vacuum in order to remove the medium. The cyanobacterial cells attached membrane is completely immersed into 1 mL

dry-ice precooled methanol in a petri-dish. Place the dish in −80 °C freezer for 30–60 min to quench cell metabolism. This is the sample for time zero (t_0).

4. Supplement the rest of the culture with labeled substrate (e.g., 5 mM of the U-^{13}C-labeled glucose) on the flask. Repeat **step 3** for a series of time points (e.g. 20 s, 60 s, 2 min, 4 min, 8 min, 16 min, 30 min, 60 min, 120 min, etc.).

3.2.2 Metabolite Extraction (Slightly Different from Subheading 3.1.2)

1. Remove samples from −80 °C freezer and place them on an icebox filled with dry ice. On each petri-dish, scrape the cells off the membrane with a cell scraper. Transfer the cell suspension into a 2-mL EP tube. Centrifuge for 1 min at 10,000 × *g* and 4 °C to pellet the cells. Transfer the supernatant into a new 1.5-mL EP tube and set aside at 4 °C.
2. Resuspend the pellet in 300 μL of extraction solution (50:50 of methanol: water) by vortexing and let it sit at 4 °C for 5 min. Centrifuge for 1 min at 10,000 × *g* and 4 °C, combine the supernatant with the supernatant obtained from **step 1**.
3. Repeat **step 2** for a final round of extraction.
4. Transfer the cell extract into HPLC vials.
5. Repeat **steps 1–4** for samples at all time points.
6. Evaporate the extract solutions under N_2 gas flow (using an N-Evap system) until dry and resuspend in 200 μL extract solution (*see* **Note 6**).

3.2.3 Metabolite Analysis by HPLC-MS/MS

Follow the procedures described in Subheading 3.1.3.

3.3 Analysis of Dynamic Labeling Data for a Metabolite

1. Calculate the fraction of unlabeled form for each metabolites in all time points sample by the formula below:

   Fraction of unlabeled metabolite A = Peak Intensity$_{\text{unlabeled isotope}}$/Sum of peak intensity$_{\text{all unlabeled, partially labeled and fully labeled isotopes}}$. *See* Fig. 4 as an example.
2. For qualitative purpose, estimate the $t_{0.5}$ for each metabolite. For example, in the case of glucose metabolism of *Synechocystis* (as shown in Fig. 4), the $t_{0.5}$ for glucose-6-phosphate/Fructose-6-phosphate is shorter than 20 s. $t_{0.5}$ for key metabolites in *Synechocystis* upon U-^{13}C-glucose labeling is listed in Table 2.
3. For quantitative purpose, calculate first-order rate constant k_a for the first metabolite downstream of substrate by fitting data to the exponential Eq. (3). Using the first-order rate constant obtained and the intracellular metabolite concentration (determined separately in Subheading 3.1.4), calculate flux through metabolite A. It equals the product of k_a and the intracellular

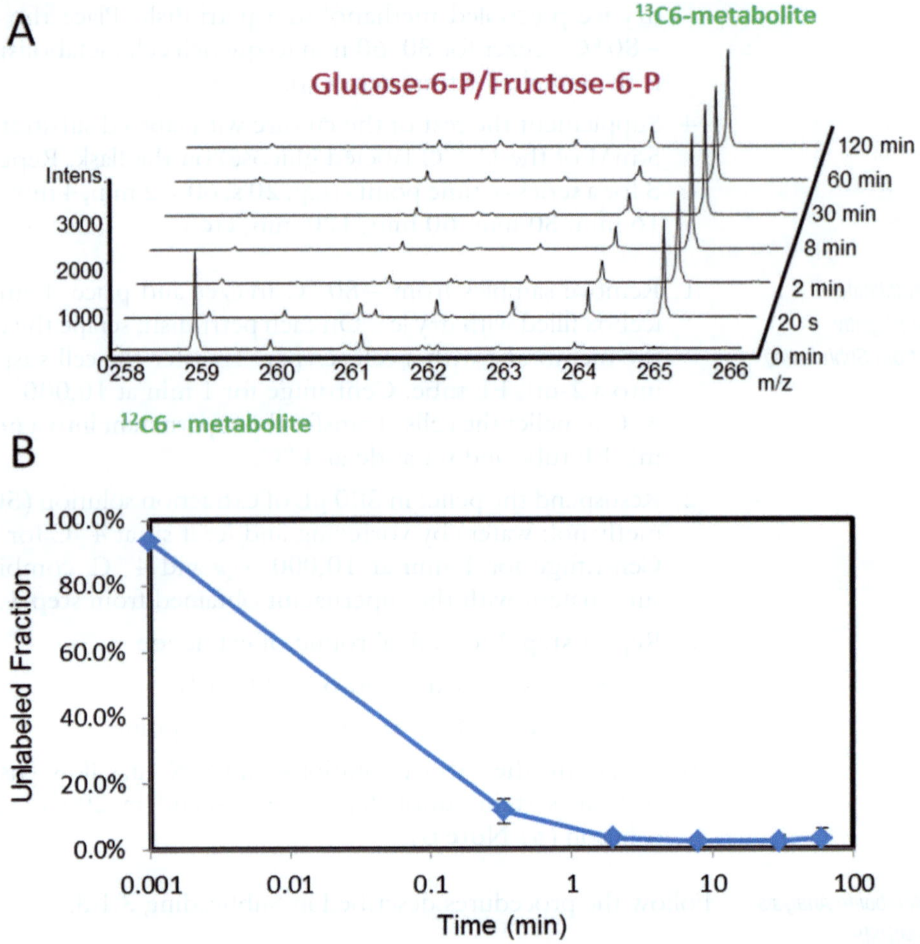

Fig. 4 Kinetic labeling pattern of glucose-6-phosphate/Fructose-6-phosphate (G6P/F6P) in *Synechocystis* cells during U-^{13}C-glucose labeling for 2 h. Unlabeled G6P/F6P (*m/z*: 259.02) are kinetically replaced by their labeled counterparts, dominantly in fully labeled forms (*m/z*: 265.04) (**a**) Raw MS spectrums for different time points; (**b**) Fraction of unlabeled G6P/F6P as a function of time

concentration of *A*, that is, $f_a = k_a \times A^T$. For metabolite B, the observed value of k_a was used in Eq. (5) and fitting kinetic labeling data of metabolite B to the Eq. (5) to obtain k_b. f_b is then calculated by $f_b = k_b \times B^T$ where B^T is determined through Subheading 3.1.4 (*see* **Note 9**).

4 Notes

1. Herein we describe an ideal status in which the first metabolite downstream of substrate can be measured. However, in many cases, the measurements of this first metabolite downstream of the substrate are not available. Consider metabolite Z, downstream of metabolite Y, where Y is not directly downstream of

Table 2
Half-labeling time ($t_{0.5}$) of key metabolites in *Synechocystis* during U-^{13}C-glucose labeling

Pathway	Metabolites	$t_{0.5}$: time for 50% label
Glycolysis/Pentose Phosphate Pathway/ Calvin Benson Cycle/ Non-mevalonate pathway	Glucose 6-phosphate	<20 s
	Fructose 6-phosphate	<20 s
	Fructose bisphosphate	<20 s
	Glyceraldehyde 3-phosphate	~20 s
	Dihydroxyacetone phosphate	~20 s
	Phosphoglycerate	~20 s
	Phosphoenolpyruvate	~20 s
	6-phosphogluconate	<20 s
	Ribulose-5-phosphate	~20 s
	Xylulose 5-phosphate	<20 s
	Ribulose 1,5-bisphosphate	<20 s
	Sedoheptulose-7-phosphate	<20 s
	Erythrose-4-phosphate	<20 s
	2-*C*-Methyl-D-erythritol 2,4-cyclodiphosphate	20 s to 2 min
Fatty Acids Synth esis	Acetyl CoA	20–30 s
	Malonyl CoA	~30 s
The TCA cycle	Fumarate	2–8 min
	Malate	2–8min
	2-oxoglutarate	30–60min
	L-Glutamic acid	30–60min
	Succinate	30–60min

the labeled substrate. In this case, the Eqs. (3) and (5) are still applicable for calculating rate constants of Y and Z, respectively, if we observe in experiment that the unlabeled form of Y satisfies single exponential decay. However, it should be noted that in this case, rate constant for Y obtained by Eq. (3) is actually **apparent rate constant** $\left(k'_y\right)$ and cannot be used to calculate flux through Y directly. Instead, the real rate constant of Y (k_y) can be derived through the apparent rate constant of its precursor (metabolite X) k'_x with Eq. (3), and then apply k'_x in Eq. (5). The k_y can be obtained by fitting kinetic labeling data of metabolite Y to Eq. (5). Briefly, to calculate the flux through a metabolite which is not the direct product of the labeled substrate, calculating the **apparent rate constant (k')** of its precursor by Eq. (3) and **the rate constant (k)** of itself by Eq. (5) is required.

2. Isotopic substrates selected for labeling experiment depend on research purposes. For example, sodium ^{13}C-bicarbonate/$^{13}CO_2$ can be used for addressing photoautotrophic metabolism of the cyanobacterium [5], uniformly or positionally labeled glucose is suitable for understanding photomixotrophic or photoheterotrophic metabolism [13, 14] and we used U-^{13}C-glutamate for tracing the TCA cycle of *Synechocystis* [8].
3. Different types of Mass Spectrometer linked to HPLC are suitable for isotope-based flux analysis such as Quadrupole Time-of-Flight (Q-TOF), orbit trap that have high mass accuracy.
4. Growth conditions of microbial cells are flexible depending on research purposes, cell types, laboratories, etc. The conditions used here should be consistent with following isotope tracer experiment.
5. 50:50 methanol/water is used as extraction solvent. Other solvents may be also useful. It has been reported that extracting specific groups of metabolites may need special solvents. For example, mixtures of acidic (0.1 M formic acid-containing) acetonitrile/water (80:20) or acetonitrile/methanol/water (40:40:20) gave rise to superior adenosine triphosphate (ATP) yields [15].
6. This step is optional. It should be noted that concentrating metabolites extract could lead to ion suppression in MS measurement and degradation of chemically unstable metabolites. This step is only necessary when the concentration of targeted metabolite is very low in the sample.
7. HPLC-MS/MS method is based on an ion-pairing chromatography first described in Luo et al. [16]. It has been validated as a universal metabolomics approach by applying standards

(Specific metabolite-dependent parameters in LC-MS *see* Table 1) and intracellular metabolite samples.

8. Two parameters described here are species-specific including average volume of a cell (V_{cell}) and cell numbers per OD (n). We strongly recommend determining these values experimentally if microbes other than *Synechocystis* are studied.
9. Although the computational procedures are beyond the scope of this chapter, it is still important to understand the mathematic approach behind DFA. For more details of computational procedures, please refer to our previous work [7]. It provided a simple but specific example dissecting nitrogen assimilation fluxes in a photosynthetic microbe.

Acknowledgments

This work was supported by the National Renewable Energy Laboratory (NREL) Director's Fellowship (Laboratory Directed Research and Development Subtask 06271403. This work was authored in part by Alliance for Sustainable Energy, LLC, the Manager and Operator of the National Renewable Energy Laboratory for the U.S. Department of Energy (DOE) under Contract No. DE-AC36-08GO28308. The views expressed in the chapter do not necessarily represent the views of the DOE or the U.S. Government. The U.S. Government retains and the publisher, by accepting the article for publication, acknowledges that the U.S. Government retains a nonexclusive, paid-up, irrevocable, worldwide license to publish or reproduce the published form of this work, or allow others to do so, for U.S. Government purposes.

References

1. Winter G, Krömer JO (2013) Fluxomics—connecting omics analysis and phenotypes. Environ Microbiol 15(7):1901–1916. https://doi.org/10.1111/1462-2920.12064
2. Orth JD, Thiele I, Palsson BO (2010) What is flux balance analysis? Nat Biotechnol 28 (3):245–248
3. Zamboni N, Fendt S-M, Ruhl M, Sauer U (2009) 13C-based metabolic flux analysis. Nat Protoc 4(6):878–892
4. Yuan J, Bennett BD, Rabinowitz JD (2008) Kinetic flux profiling for quantitation of cellular metabolic fluxes. Nat Protoc 3 (8):1328–1340
5. Young JD, Shastri AA, Stephanopoulos G, Morgan JA (2011) Mapping photoautotrophic metabolism with isotopically nonstationary ^{13}C flux analysis. Metab Eng 13(6):656–665
6. Yuan J, Fowler WU, Kimball E, Lu W, Rabinowitz JD (2006) Kinetic flux profiling of nitrogen assimilation in *Escherichia coli*. Nat Chem Biol 2(10):529–530
7. Wu C, Xiong W, Dai J, Wu Q (2016) Kinetic flux profiling dissects nitrogen utilization pathways in the oleaginous green alga *Chlorella protothecoides*. J Phycol 52(1):116–124. https://doi.org/10.1111/jpy.12374
8. Xiong W, Brune D, Vermaas WFJ (2014) The γ-aminobutyric acid shunt contributes to closing the tricarboxylic acid cycle in *Synechocystis* sp. PCC 6803. Mol Microbiol 93(4):786–796. https://doi.org/10.1111/mmi.12699
9. Xiong W, Morgan JA, Ungerer J, Wang B, Maness P-C, Yu J (2015) The plasticity of cyanobacterial metabolism supports direct CO_2

conversion to ethylene. Nat Plants 1:15053. https://doi.org/10.1038/nplants.2015.53

10. Gao X, Gao F, Liu D, Zhang H, Nie X, Yang C (2016) Engineering the methylerythritol phosphate pathway in cyanobacteria for photosynthetic isoprene production from CO_2. Energy Environ Sci 9(4):1400–1411. https://doi.org/10.1039/C5EE03102H
11. Amador-Noguez D, Feng X-J, Fan J, Roquet N, Rabitz H, Rabinowitz JD (2010) Systems-level metabolic flux profiling elucidates a complete, bifurcated tricarboxylic acid cycle in clostridium acetobutylicum. J Bacteriol 192(17):4452–4461. https://doi.org/10.1128/jb.00490-10
12. Young JD (2014) INCA: a computational platform for isotopically non-stationary metabolic flux analysis. Bioinformatics 30 (9):1333–1335. https://doi.org/10.1093/bioinformatics/btu015
13. You L, Berla B, He L, Pakrasi HB, Tang YJ (2014) ^{13}C-MFA delineates the photomixotrophic metabolism of *Synechocystis* sp. PCC 6803 under light- and carbon-sufficient conditions. Biotechnol J 9 (5):684–692. https://doi.org/10.1002/biot.201300477
14. You L, He L, Tang YJ (2015) Photoheterotrophic fluxome in *Synechocystis* sp. strain PCC 6803 and its implications for cyanobacterial bioenergetics. J Bacteriol 197(5):943–950. https://doi.org/10.1128/jb.02149-14
15. Rabinowitz JD, Kimball E (2007) Acidic acetonitrile for cellular metabolome extraction from *Escherichia coli*. Anal Chem 79 (16):6167–6173. https://doi.org/10.1021/ac070470c
16. Luo B, Groenke K, Takors R, Wandrey C, Oldiges M (2007) Simultaneous determination of multiple intracellular metabolites in glycolysis, pentose phosphate pathway and tricarboxylic acid cycle by liquid chromatography–mass spectrometry. J Chromatogr A 1147 (2):153–164

Chapter 15

Network Modeling of Complex Data Sets

Piet Jones, Deborah Weighill, Manesh Shah, Sharlee Climer, Jeremy Schmutz, Avinash Sreedasyam, Gerald Tuskan, and Daniel Jacobson

Abstract

We demonstrate a selection of network and machine learning techniques useful in the analysis of complex datasets, including 2-way similarity networks, Markov clustering, enrichment statistical networks, FCROS differential analysis, and random forests. We demonstrate each of these techniques on the *Populus trichocarpa* gene expression atlas.

Key words Differential analysis, FCROS, Fisher exact test, Enrichment, Similarity network, Random forests, Machine learning

1 Introduction

1.1 Basic Network Theory

Networks are useful tools for the representation and analysis of complex biological datasets. A single network represents a system as a collection of objects (nodes) connected by links (edges) representing relationships between the objects [1] (*see* Fig. 1A). A node can represent any biological object (gene, protein, sample, phenotype, metabolite, etc.), and edges can represent any qualitative or quantitative relationship between pairs of objects, for example the co-expression between genes, or the similarity between the microbial species content of two soil samples.

This manuscript has been authored by UT-Battelle, LLC under Contract No. DE-AC05-00OR22725 with the U.S. Department of Energy. The United States Government retains and the publisher, by accepting the article for publication, acknowledges that the United States Government retains a non-exclusive, paid-up, irrevocable, worldwide license to publish or reproduce the published form of this manuscript, or allow others to do so, for United States Government purposes. The Department of Energy will provide public access to these results of federally sponsored research in accordance with the DOE Public Access Plan (http://energy.gov/downloads/doe-public-access-plan). The work conducted by the U.S. Department of Energy Joint Genome Institute is supported by the Office of Science of the U.S. Department of Energy under Contract No. DE-AC02-05CH11231.

The authors Piet Jones and Deborah Weighill contributed equally to this work.

Michael E. Himmel and Yannick J. Bomble (eds.), *Metabolic Pathway Engineering*, Methods in Molecular Biology, vol. 2096, https://doi.org/10.1007/978-1-0716-0195-2_15, © Springer Science+Business Media, LLC, part of Springer Nature 2020

A

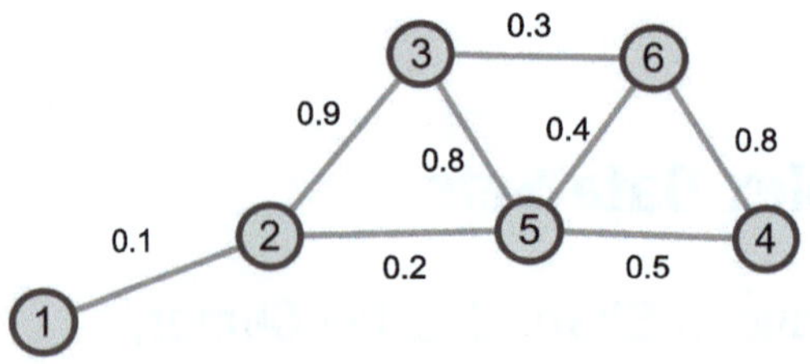

B

	1	2	3	4	5	6
1	0	0.1	0	0	0	0
2	0.1	0	0.9	0	0.2	0
3	0	0.9	0	0	0.8	0.3
4	0	0	0	0	0.5	0.8
5	0	0.2	0.8	0.5	0	0.4
6	0	0	0.3	0.8	0.4	0

C

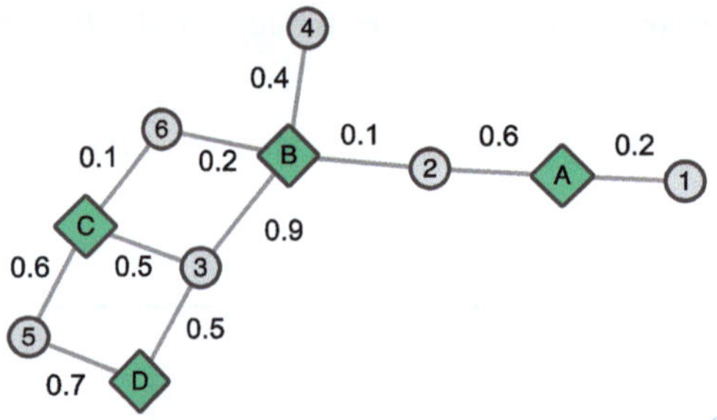

D

	A	B	C	D
1	0.2	0	0	0
2	0.6	0.1	0	0
3	0	0.9	0.5	0.5
4	0	0.4	0	0
5	0	0	0.6	0.7
6	0	0.2	0.1	0

Fig. 1 Networks and adjacency matrices. (**A**) Network (**B**) Adjacency Matrix, respective entries in the adjacency matrix indicate the edge weight of an edge that is between nodes which are given by the respective row/column pairs. A weight of zero indicates the absence of an edge, while in an unweighted network the default edge weight is 1. (**C**) Bipartite Network (**D**) Adjacency matrix for bipartite network

Networks in which each node is of the same type can be represented by a standard adjacency matrix, which is simply a table in which rows and columns represent nodes, and each entry *ij* is 1 if node *i* is connected to node *j*, or zero otherwise [2]. Alternatively, one can set each entry *ij* to the strength of the relationship between nodes *i* and *j*, called the *edge weight* [2]. For example, nodes can represent genes within an organism, and edges can represent the co-expression between those genes across various tissues. A simple example of a network and its adjacency matrix are shown in Fig. 1A and B, respectively.

One can also have networks in which the nodes fall into one of two classes. Such networks are called bipartite networks [2]. For example, each node in a bipartite network can represent either a sample or a species, and we connect a species node to a sample node if the species occurs within that sample. A small example bipartite network and its adjacency matrix are shown in Fig. 1C and D, respectively.

1.2 Data Matrix and Overview

Data can usually be structured into a matrix (table) in which columns represent samples and rows represent variables measured across the samples. Thinking abstractly about such a data structure, there are various ways in which we can probe the data. We can compare the rows pairwise in order to gain understanding of the relationships between variables, we can compare the columns pairwise in order to gain understanding of the relationships between samples, we can relate particular rows to particular columns in order to understand which variables are particularly important in which samples and we can perform differential analysis to identify which variables have significant differences across samples.

In this chapter, we describe a protocol for the probing and unpacking of a complex biological dataset which involves each of these different types of analyses, and we demonstrate the outcomes of these analyses on the *Populus trichocarpa* gene expression atlas dataset. In particular we will outline approaches for the following:

Enrichment Networks: Sample-variable enrichment allows for the statistical association between samples and variables. Given a matrix in which rows represent variables and columns represent samples, the right-tailed Fisher exact test can be used to answer the question "is variable *x* enriched in sample *y*". This is similar to the concept of determining if an ontology term is enriched in a set of genes, as done by Gene Ontology enrichment software such as GOEAST [3]. Applying matrix-wide Fisher exact tests to the gene atlas expression matrix will determine which samples each gene's expression is enriched in.

Sample Similarity, Gene Co-expression and Clustering: Pairwise comparison of the expression profiles of genes across tissues and subsequent thresholding allows for the construction of a gene co-expression network (see for example [4, 5]). Clustering of the co-expression network with a clustering algorithm such as Markov Clustering (MCL) [6, 7] allows one to extract groups of genes which have similar expression relationships across tissues. One can also perform a pairwise comparison of sample vectors to identify which samples have similar overall gene expression patterns.

DUO Similarity Networks: DUO is a similarity metric developed by Sharlee Climer [8]. It categorizes values in an expression matrix into high, medium and low values, and then for each pair of objects, it calculates a scaled co-occurrence of all four possible combinations of high values and low values. Thus, unlike most similarity metrics, the comparison of the expression profiles of two genes will result in four comparisons: high values in gene A vs high values in gene B, high values in gene A vs low values in gene B, low values in gene A vs high values in gene B, and low values in gene A vs low values in gene B. The structure of the DUO similarity metric is similar to the SNP (Single Nucleotide Polymorphism) correlation metric, CCC, also developed by Sharlee Climer et al. [9, 10].

FCROS Differential Analysis: We can investigate the difference between parts of a variable vector, where the variable vector is partitioned based on some sample characterization. The Fold Change Rank Order Statistic (FCROS), from [11], is a method that allows us to obtain statistically significant estimates of the degree of difference of observations (variables) measured under a set of conditions. In our case we can view sets of samples (columns), as representing these respective conditions, and can therefore answer the question of whether or not a variable's measurements are higher/lower in one condition versus another.

Random Forests: Random Forest is a machine learning method [12]. It uses an ensemble approach for classification and or regression. This involves the construction and *growing* of multiple decision trees followed by the aggregation of the results from the respective trees. It is therefore essentially a bagging approach. The out-of-bag error for each variable, over all trees, can be used to assess the respective variable's importance in the overall classification or regression. It is worth noting that this approach is robust to overfitting and noisy data [13].

2 Materials

2.1 Data

The *Populus trichocarpa* gene atlas expression dataset was analyzed on the Oak Ridge National Laboratory's OLCF (Oak Ridge Leadership Computing Facility) supercomputer platform. The RNA-Seq dataset consisted of Illumina paired-end sequencing reads for 81 samples representing various tissues, including roots, root tips, buds, stems, leaves, etc. at different growth and developmental stages.

2.2 Software/Packages

Table 1 provides a list of publicly available software packages/libraries used in this analysis. Basic custom Perl and R scripts were also used.

3 Methods

3.1 Data Preparation

1. Quality trim the paired end reads using Skewer software [24].
2. Map the reads to the *Populus trichocarpa* reference genome using STAR [25] (*see* **Note 1**).
3. Calculate transcripts per kilobase million (TPMs), as follows:
   (a) Divide the read counts by the length of the gene in kilobases. This gives the reads per kilobase (RPK).

Table 1
Packages, libraries and resources used

Resource	Reference
Perl libraries	Available from http://www.cpan.org/
Text::NSP::Measures::2D::Fisher Perl module	[14]
Graph::Undirected	Jarkko Hietaniemi
R libraries/resources	
R	[15]
RStudio	[16]
data.table	[17]
fcros	[11]
ggplot2	[18]
ggthemes	[19]
pbdMPI	[20]
randomforest	[21]
reshape	[22]
Other	
MCL-Edge	[6, 7]
Cytoscape	[23]

(b) Count up all the RPK values in a sample and divide this number by 1,000,000. This gives the "per million" scaling factor.

(c) Divide the RPK values by the "per million" scaling factor. This gives the TPM.

3.2 Matrix-Wide Fisher Exact Test and Enrichment Networks

1. Construct the $m \times n$ gene expression matrix M in which rows represent genes, columns represent samples, and each entry represents the expression (TPM) value of gene i in tissue j.
2. Scale each entry x_{ij} in each row i as follows: $x_{ij}^* = \mathrm{int}(\frac{x_{ij}}{\sum_j x_{ij}} \times 100)$ (*see* **Note 2**).
3. For each entry x_{ij} in the matrix M, construct the contingency table shown in Fig. 2B and calculate the right-tailed Fisher exact test on that contingency table (*see* **Note 3**).
4. Perform the Benjamini-Hochberg procedure [26] for controlling the false discovery rate across all $N = m \times np$-values in order to determine which sample-variable associations are

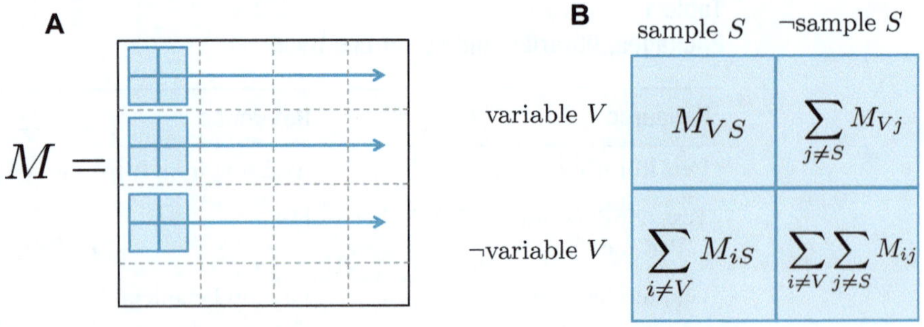

Fig. 2 Matrix Fisher exact test. (**A**) The Fisher exact test is calculated for every entry in the expression matrix *M*. (**B**) Contingency table constructed for each entry in the expression matrix *M*

statistically significant as follows: Let P_1, P_2, $\ldots P_N$ be the $N p$-values in ascending order, and their corresponding hypotheses H_1, H_2, …, H_N. Sequentially for each *p*-value, determine whether $P_i \leq \frac{i}{N}\alpha$ where $i \in \{1 \ldots N\}$ for a chosen FDR level of α. Find the largest $i = L$ for which the inequality holds, and reject null hypotheses $H_1 \ldots H_L$ (*see* **Note 4**). If there are *p*-values which tie with the *p*-value of the last rejected hypothesis, we reject those as well.

5. Represent the resulting rejected hypotheses (associations between genes and samples) as a bipartite network in which each node represents either a gene or a sample, and each edge represents a rejected null hypothesis, statistically associating a gene with a sample, in `sif` format (*see* **Note 5**).
6. Visualize the resulting network by loading the `sif` file into Cytoscape [23] and applying the desired color schemes (*see* **Note 6**, Fig. 3).

3.3 Variable Similarity Networks and Clustering

1. Construct the $m \times n$ gene expression matrix M in which rows represent genes, columns represent samples, and each entry represents the expression (TPM) value of gene i in tissue j.
2. Calculate the Pearson correlation coefficient between all pairs of genes (rows) using the mcxarray program from the MCL-Edge software package [6, 7] available from http://micans.org/mcl/ (*see* **Notes 7** and **8**).
3. Convert the resulting output in mcl matrix format to line-based format using the mcxdump program in the MCL-Edge software package, and then convert the line-based format to `sif` format (*see* **Note 9**).
4. Choose a threshold t, and retain only edges (lines in the `sif` file) for which $|w| \geq t$ where w is the Pearson correlation edge weight (*see* **Note 10**).

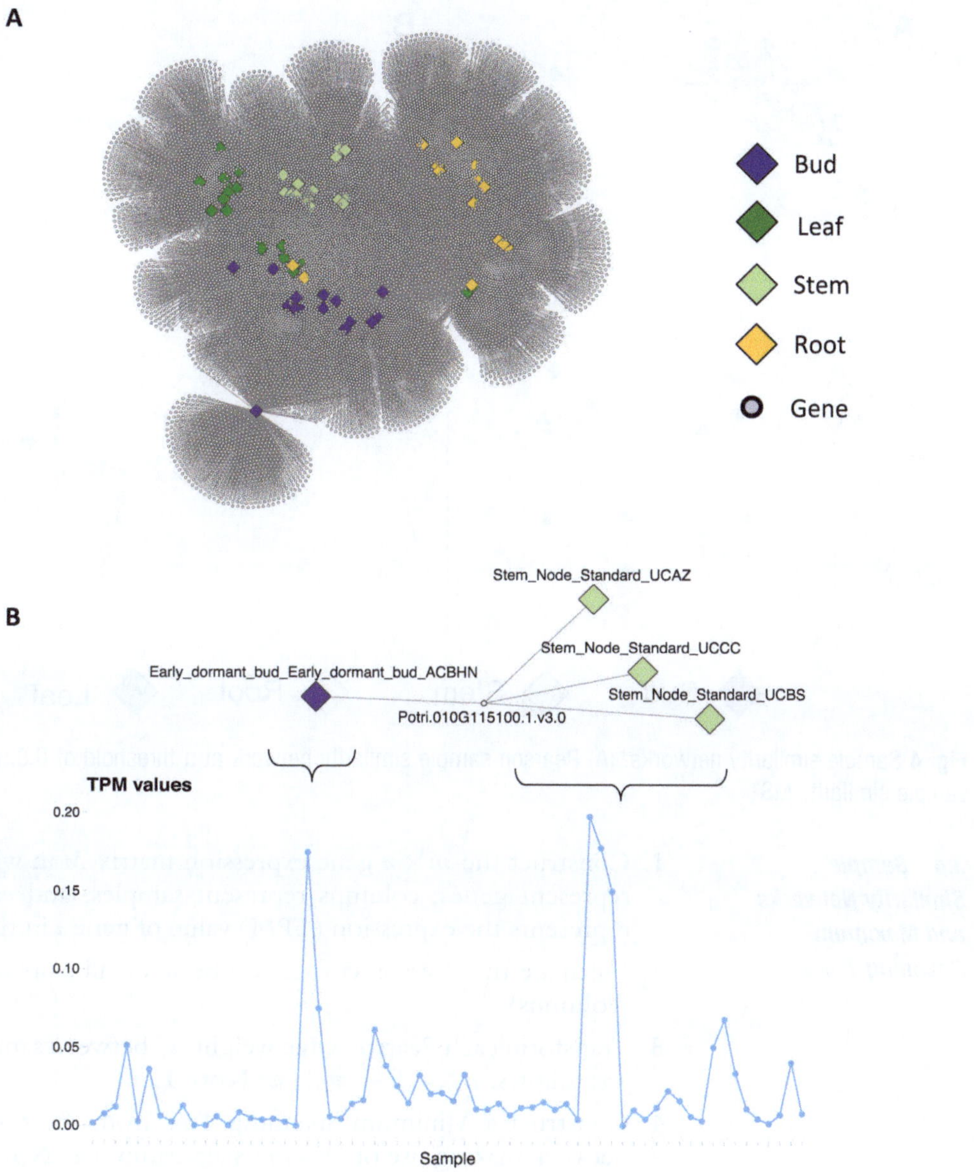

Fig. 3 Enrichment network. (**A**) Enrichment network for the *P. trichocarpa* gene atlas. Large diamond nodes represent samples, colored according to source tissue. Small grey nodes represent genes. An edge connects a gene to a sample if that gene's expression is significantly expressed in that sample, as determined using the right-tailed Fisher exact test and FDR correction. (**B**) An example of a gene enriched in 4 of the samples and the expression profile (TPM values) of the gene

5. Load the `sif` file into Cytoscape for visualization.
6. Use MCL [6, 7] to cluster the thresholded similarity network into modules of co-expressed genes (*see* **Note 11**).

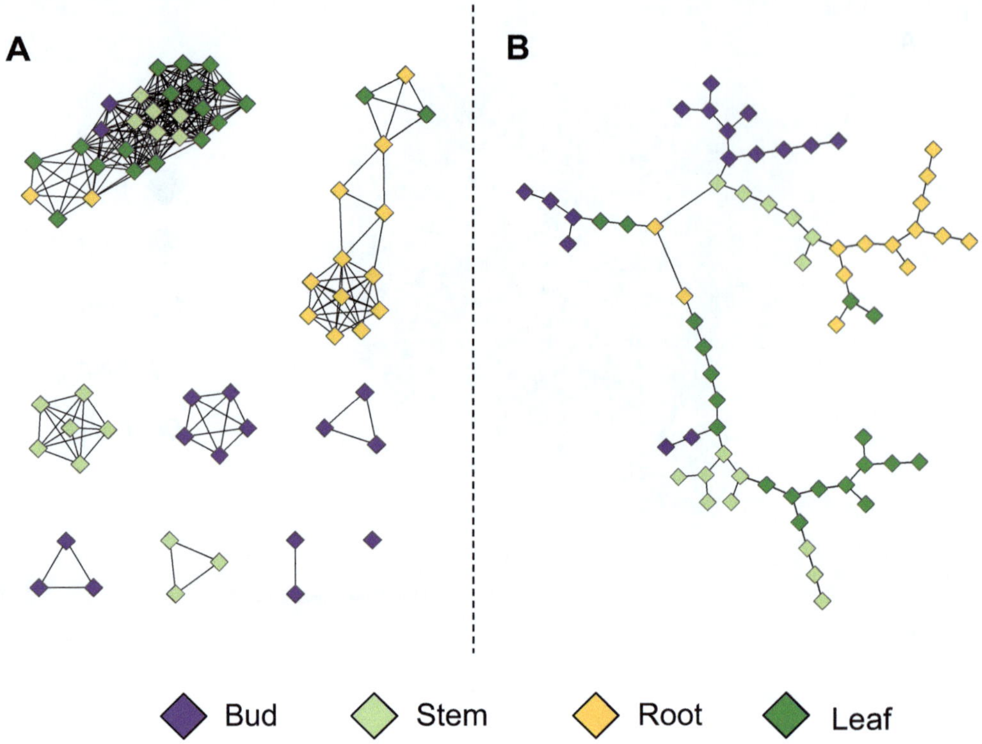

Fig. 4 Sample similarity networks. (**A**) Pearson sample similarity network at a threshold of 0.8. (**B**) Pearson sample similarity MST

3.4 Sample Similarity Networks and Maximum Spanning Trees

1. Construct the $m \times n$ gene expression matrix M in which rows represent genes, columns represent samples, and each entry represents the expression (TPM) value of gene i in tissue j.
2. Calculate the Pearson correlation between all pairs of samples (columns).
3. Transform each Pearson edge weight w_{ij} between samples i and j as follows: $w_{ij}^{*} = 1 - |w_{ij}|$ (*see* **Note 12**).
4. Construct a Minimum Spanning Tree from the transformed network making use of Dijkstra's algorithm (*see* **Note 13**).
5. Visualize the network in Cytoscape (*see* Fig. 4).

3.5 DUO Similarity Networks

1. Given our expression matrix M in which rows represent genes and columns represent samples, scale each entry x_{ij} as follows: $x_{ij}^{*} = \frac{x_{ij}}{\max_i(x_{ij})}$ (*see* **Note 14**).
2. Determine upper and lower thresholds U and L, respectively, such that 25% of the values in the scaled expression matrix lie above U and 25% of the values in the scaled expression matrix lie below L. Values above the upper threshold are marked as "high", values below the lower threshold are marked as "low", and the remaining values are marked as neutral.

3. For each pair of genes A and B, denote the high values of A and B as A_H and B_H, respectively, and the low values of A and B as A_L and B_L, respectively. For each pair ij where $i \in \{A_H, A_L\}$ and $j \in \{B_H, B_L\}$, calculate the DUO similarity metric as $\mathrm{DUO}_{ij} = 4D_{ij}(1 - \frac{f_i}{1.5})(1 - \frac{f_j}{1.5})$ where D_{ij} represents the fraction of the vector length in which i and j co-occur, f_i and f_j represent the fraction of i and j in genes A and B, respectively (*see* **Note 15**).
4. Convert the resulting DUO network to `sif` format.
5. Threshold the resulting DUO network (this analysis used a threshold of 0.8), represent the network in `sif` format and load it into Cytoscape for visualization (*see* Fig. 5, **Note 16**).

3.6 Fold Change Rank Order Statistic Differential Analysis

1. Load an expression matrix into R, here the rows represent genes, columns represent samples, and the values represent expression values (TPM) of the given gene in the respective sample (*see* **Note 17**).
2. Create a model (design) matrix, where the original samples (columns of the expression matrix) are the rows, and the sample groups that will be compared are the columns. Values in the matrix are binary, indicating association between individual samples and sample groups, respectively (*see* **Note 18**).
3. Use voom [27] to perform mean-variance stabilization of the expression matrix, given the model matrix (*see* **Note 19**).
4. Determine valid pairs of sample groups for the differential analysis (*see* **Note 20**).
5. For each of the respective pairs, sub-select the expression matrix that only contains those respective columns (samples), obtaining a list of sub-matrices.
6. Run the fcros function on the respective sub-matrices, assigning one sample group in the pair as the control, and the other as the case (*see* **Note 21**).
7. Filter the respective results by p-value (say, for an $\alpha = 0.01$). Remove results that have a f-value between a given probability bound (say, $0.1 <$ f-value < 0.9). Report for each gene the sample group labels used in the comparison and the $\log_2$ transform of the applicable robust fold change estimate, in a 3-column tab delimited format (*see* **Note 22**).
8. Visualize the result as a network in Cytoscape, where nodes are given by the comparison labels and genes, respectively (*see* **Note 23**).

3.7 Random Forests Sample Importance

1. Load an expression matrix into R, here the rows represents genes, columns represent samples, and the values represent expression values (TPM) of the given gene in the respective

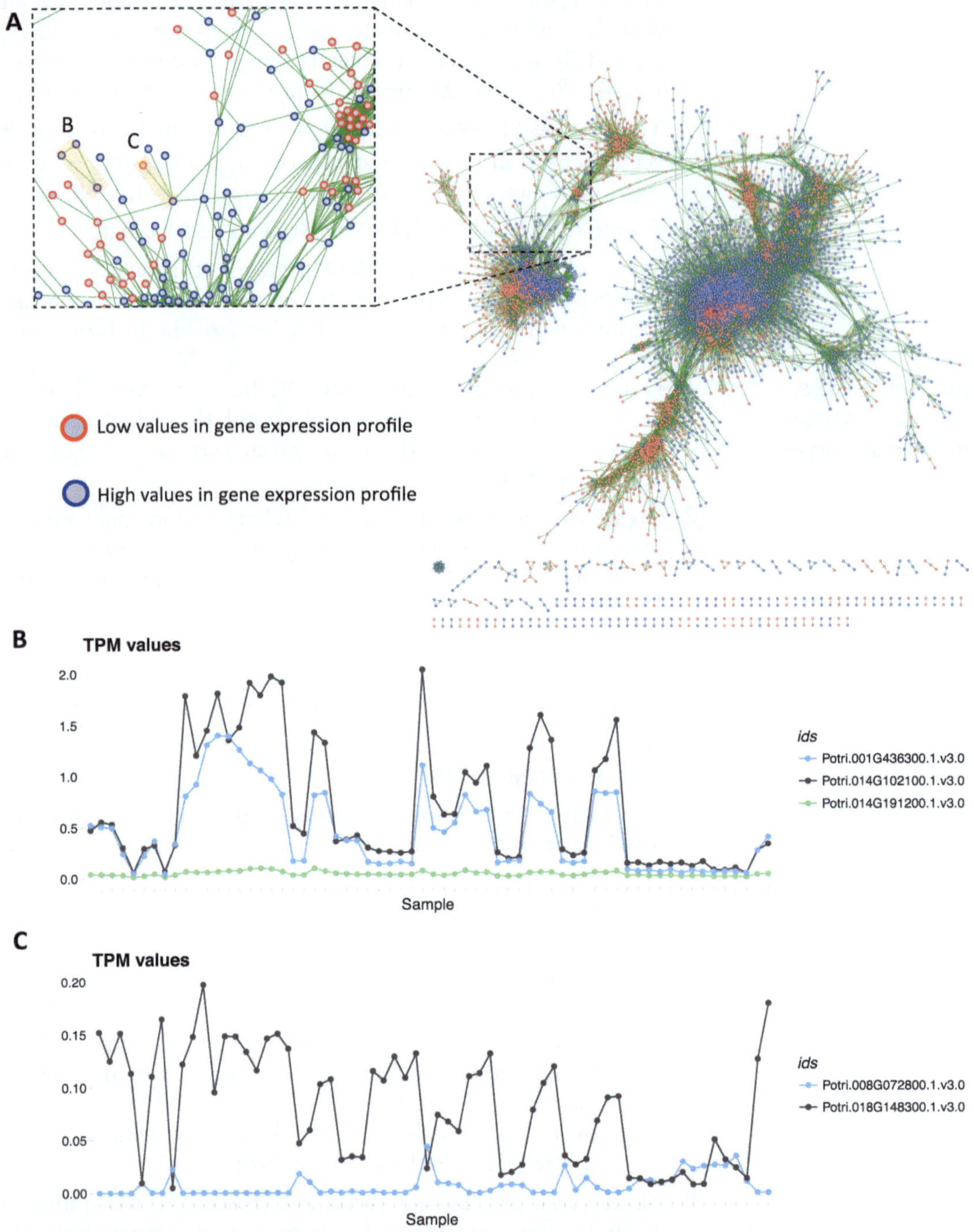

Fig. 5 DUO network. (**A**) DUO co-expression network for the *P. trichocarpa* gene expression atlas. Blue bordered nodes represent high expression values for a given gene, red-bordered nodes represent low expression values for a given gene. An edge (green) between two nodes represents the co-occurrence between the expression values of the two genes it connects, calculated using the DUO metric. (**B**) Expression profiles (TPM values) for the three genes highlighted B in panel **a**. (**c**) Expression profiles (TPM values) for the two genes highlighted C in panel **A**

sample (*see* **Note 17**). Transpose the data, thus making the samples the rows and the genes the columns.

2. From meta-data regarding the samples, create meaningful groupings. This can be as simple as grouping replicates, or more involved such as grouping by tissue.
3. For each possible paired combination of groups, create a subset of the data. Filter out all rows that do not belong to the respective groups and add a label column, containing the respective group labels (*see* **Note 24**).
4. Use the *randomForest* function from the randomForest package in R [21], ensure that the number of trees is larger than the number of columns of the data subset. Set the importance flag to TRUE (*see* **Note 25**).
5. Extract the variable importance from the resulting object by using the *importance* function and convert it to a DataFrame or data.table.
6. Sort by the *MeanDecreaseInAccuracy* column in a descending order. Plot the results and choose an appropriate cutoff. Alternatively select the top *N* genes based on a high MeanDecreaseInAccuracy value (*see* **Note 26**).
7. Collect the results from all sample pairs, recording which sample comparison was performed. Save the results in sif format (*see* **Note 27**).
8. Visualize the result as a network in Cytoscape, where nodes are given by the comparison group labels and genes, respectively (*see* **Note 28**).

4 Notes

1. The reference sequence that we used for alignment purposes was version 3 of the *Populus Trichocarpa* genome, [28], which was obtained from Phytozome [29].
2. The standard Fisher exact test is defined for integer values. We scale the expression values of each gene to take an integer value between 0 and 100.
3. The Fisher exact test can be calculated on a contingency table using a Perl script making use of the Text::NSP::Measures::2D::Fisher Perl module [14] available from the Comprehensive Perl Archive Network (CPAN) at http://search.cpan.org/dist/Text-NSP/lib/Text/NSP/Measures/2D/Fisher.pm.
4. Applying the FDR procedure can be seen as a thresholding procedure for the raw bipartite network connecting all genes to all samples, maintaining only edges which show some level of enrichment. Each rejected null hypothesis becomes an edge

in the resulting network, connecting a gene to a sample. A similar approach was used previously by Weighill and Jacobson [30].

5. `sif` format is a 3-column table format in which the first and third columns represent nodes, the second column represents an edge annotation. In the case of the enrichment network, the first column will consist of genes, the third column will consist of samples, and the second column will represent the *p*-value of the association between the sample and gene in question. The number of lines in the `sif` file should be equal to the number of rejected hypotheses (number of edges in the network).

6. Nodes and edges in Cytoscape [23] can be formatted with various visual attributes. In this example, we color the sample nodes according to tissue type. Figure 3 shows the resulting significant associations between genes and samples, represented as a bipartite network. Large, diamond nodes represent samples, small grey nodes represent genes, and each edge represents a significant enrichment of the expression of a particular gene in a particular sample. Figure 3B shows an example of a gene which is enriched in four samples—one bud sample and three stem samples. The line plot shows the TPM values for this gene across all samples, and clearly indicates the enrichment pattern which is recognized by the Fisher exact test. Line plots were constructed using R and various R resources [15, 16, 18, 19, 22].

7. Co-expression networks are widely used and can be seen in various publications. For example, *see* ref. 4, 5.

8. Pearson is one example of a similarity metric which measures the extent to which two variables co-vary. Other similarity metrics can be used to construct similarity networks, a selection of which are discussed in reference [31] and various others [32, 33]. When the data contains missing values in the form of "NA" entries, we recommend using the R *cor* function for calculating correlation coefficients, as it has a variety of options for the handling of missing values.

9. The mcxarray program outputs the resulting correlation matrix in MCL matrix format. The mcxdump program converts this to a line-based format, with one edge per line. The first two columns represent the source and target nodes, respectively, and the third column represents the edge weight (in this case Pearson correlation value). Converting to `sif` format simply involves swapping the second and third column of the line-based format.

10. The Pearson correlation coefficient produces similarity values between -1 and 1. A Pearson correlation of 1 means that the two vectors follow the same pattern of variation (i.e. when the

one vector increases, the other one increases). A Pearson correlation of − 1 means that the two vectors have opposite patterns of variation, in that if one vector increases, the other decreases. A Pearson correlation of 0 means that there is no association between the two vectors. In many cases, such as gene co-expression, one is interested in large positive and large negative Pearson correlation values. Thus we apply an absolute threshold, keeping edges for which the absolute value of the Pearson correlation is greater than a set threshold. This analysis used an absolute threshold of 0.8.

11. MCL clusters a similarity network into modules of similar nodes. In this case, where edges represent the similarity (Pearson correlation) between the expression profiles of genes, MCL clusters the co-expression network into groups of co-expressed genes. MCL requires an inflation parameter, which controls the granularity of clusters produced. A high inflation value produces a larger number of smaller clusters, whereas a low inflation parameter produces a smaller number of larger clusters [6, 7]. This analysis used an inflation value of 2. The output of MCL consists of a multi-line file. Each line is a list of genes in a particular cluster, with one line per cluster.
12. This transformation converts the Pearson correlations from a similarity measure to a distance measure.
13. Applying a minimum spanning tree algorithm to the transformed edges, and then replacing the original edge weights gives us a maximum spanning tree. Construction of the MST was done with a custom Perl script which made use of Dijkstra's algorithm in the Graph Perl module (Jarkko Hietaniemi, http://www.cpan.org/). The Perl script outputs the MST in `sif` format.
14. This transformation scales each value x_{ij} in M by dividing it by the maximum value in the row, and thus scales the values in the matrix to range between 0 and 1. This also forces the genes to vary on the same scale, and thus gives each gene an equal chance to obtain high and low values in the next step.
15. Intuitively, the DUO metric is a form of correlation/similarity metric between two manifestations of each gene—high and low. For example, a high DUO value might tell you that the high values of gene A co-occur with the low values of gene B, or that the low values of gene C co-occur with the low values of gene D. The DUO metric also scales the resulting values according to the fraction of high/low values in the vectors being compared in order to adjust the values to account for the effect of frequency. DUO outputs networks in `gml` format.

16. Figure 5A shows the resulting DUO co-expression network for the *P. trichocarpa* gene atlas, visualized in Cytoscape. Each gene is represented by two nodes, high (blue) and low (red), respectively. For example, an edge between a blue-bordered gene *a* and a red-bordered gene *b* means that the high values of gene *a* co-occur with the low values of gene *b*. Figure 5B and C show the expression profiles of the marked genes in the DUO network. One can clearly see the co-occurrence between high values in Fig. 5B and the co-occurrence between high and low values in Fig. 5C. Line plots were constructed using R and various R resources [15, 16, 18, 19, 22].
17. Here we use transcripts per million (TPM) as we are interested in modeling relative abundance. Alternatively, if a TPM matrix is not available a raw count matrix of expression values can be used, after applying TMM normalization from the edgeR package [34] to the raw count matrix.
18. The sample groups can be any biological significant grouping of samples. In most cases a sample group will be a grouping of biological replicates of a given sample. It is important that these sample groups consist of a large enough number of samples.
19. The voom adjustment procedure is used to allow methods that were originally developed for microarray data, to be applicable to RNA-seq data. It is important to note that the output of the procedure is $\log_2$ values.
20. We define valid pairs as those that are of biological interest given the experiment. This is classically built into the sampling design and should be apparent from the sample labels. Though there may be some sample group pairs that may not make sense to compare, such as Root with Nitrogen treatment vs Mature Leaf. FCROS assumes a case *versus* control comparison, where control is a reference sample group. Therefore one member of the comparison pair is the case, while the other serves as the control. This choice of case/control group can be done randomly or more in line with the sample design. This will impact the interpretation of estimates such as fold change and *f*-value (described in *see* **Note 22**).
21. When applying the fcros function it is important to set the appropriate option to indicate to the function that the input is log-transformed, as the results from the voom adjustments are $\log_2$ values.
22. Here the *f*-value is an estimate of the probability of over/under expression. Values closer to 1 for a given comparison indicates that the gene has a higher probability of over expression in the case. Alternatively, a value closer to 0 indicates a higher

probability of under expression in the case given the comparison. The element of the sample group pair that is assigned as the case or control will determine the interpretation of the *f*-value.

23. In the network visualization we have two different types of nodes, gene, and comparison, respectively. The comparison nodes summarize the case and control that was used to test for differential genes. Thus a gene node is connected to a comparison node if it was found significantly over/under expressed in that given comparison. The edges can be colored to indicate if the gene is either over expressed (say, red) or under expressed (say, blue). This information is encoded in the log value of the robust fold change estimate. A negative value indicates under expression, while a positive value indicates over expression. Furthermore, the log fold change can be used to weight the edges, with higher absolute values representing a darker shade and lower values a lighter shade. As an example *see* Fig. 6.
24. This way of generating groups answer the question: which genes best discriminate between these two respective groups? Instead of taking all pairs of groups, we can also assign the same label to members that are not in a target group. This answers the question: which genes best discriminate between our target group and everything else?
25. The default parameters, apart from the number of trees and the *importance* parameters, should be sufficient for most datasets.
26. The *MeanDecreaseInAccuracy* is one measure of the contribution that a given gene has to the classification accuracy, should that gene be removed. There are alternative measures that can be used, each answering a slightly different hypothesis. For our purposes the *MeanDecreaseInAccuracy* is more appropriate. Plotting the curve allows us to identify an appropriate threshold.
27. The sif format is preferred as it is easy to important into Cytoscape. In this case we should have the first column as the genes that were deemed important based on the cutoff, the second column should be the *MeanDecreaseInAccuracy* value, and the last column should be the name of the two groups that were compared.
28. After loading the network into Cytoscape, assigning different colors to different classes of variables or nodes will make it easier to uncover patterns. An example can be seen in Fig. 7.

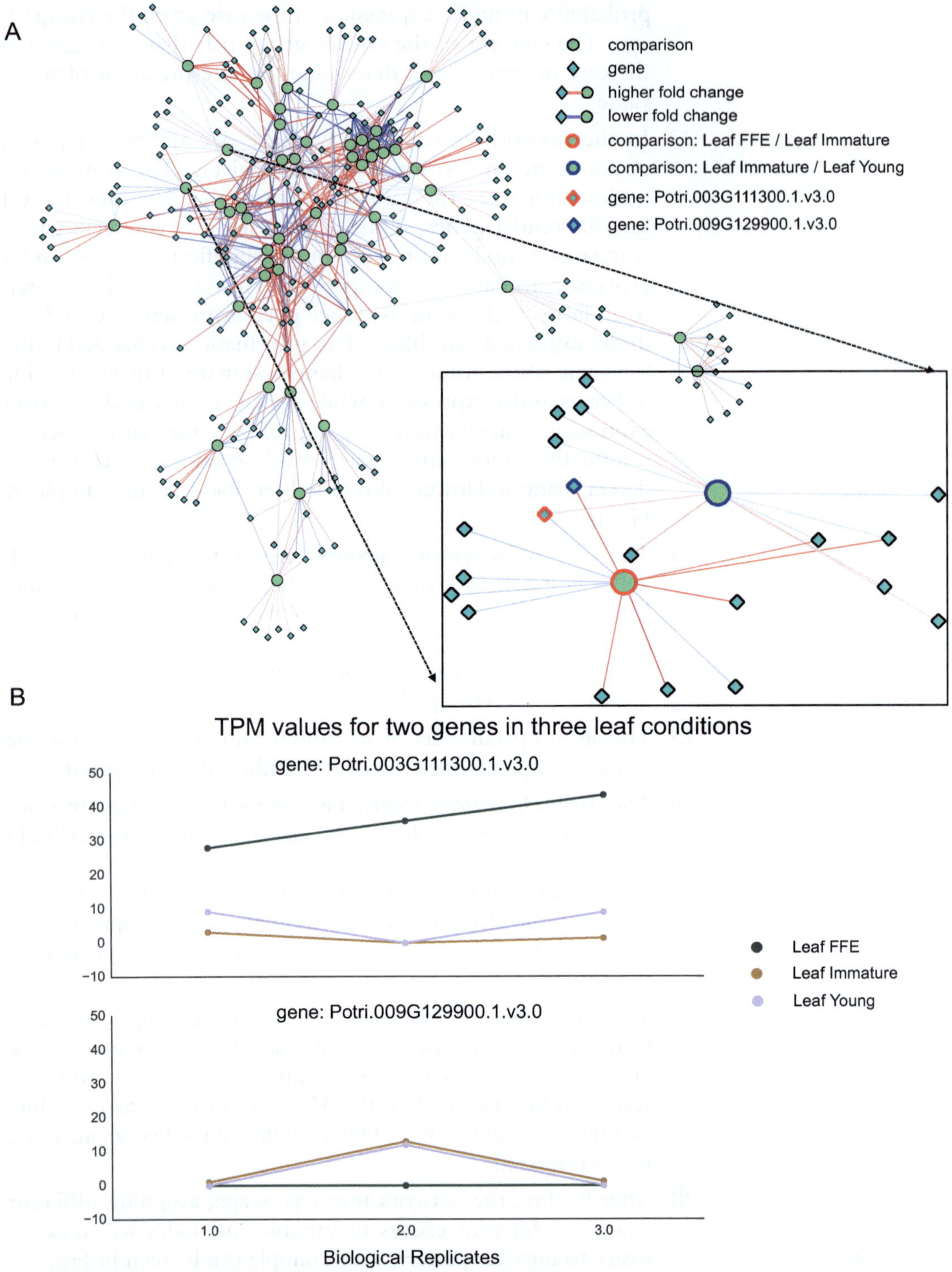

Fig. 6 FCROS network. (**A**) Network visualization of the differential results from the *P. trichocarpa* gene atlas gene expression data. Round light green nodes indicate comparisons between sample groups against which differential genes, the dark green diamond nodes, were tested. An edge between a gene node and a comparison node indicates that the given gene was significantly differentially expressed in the respective

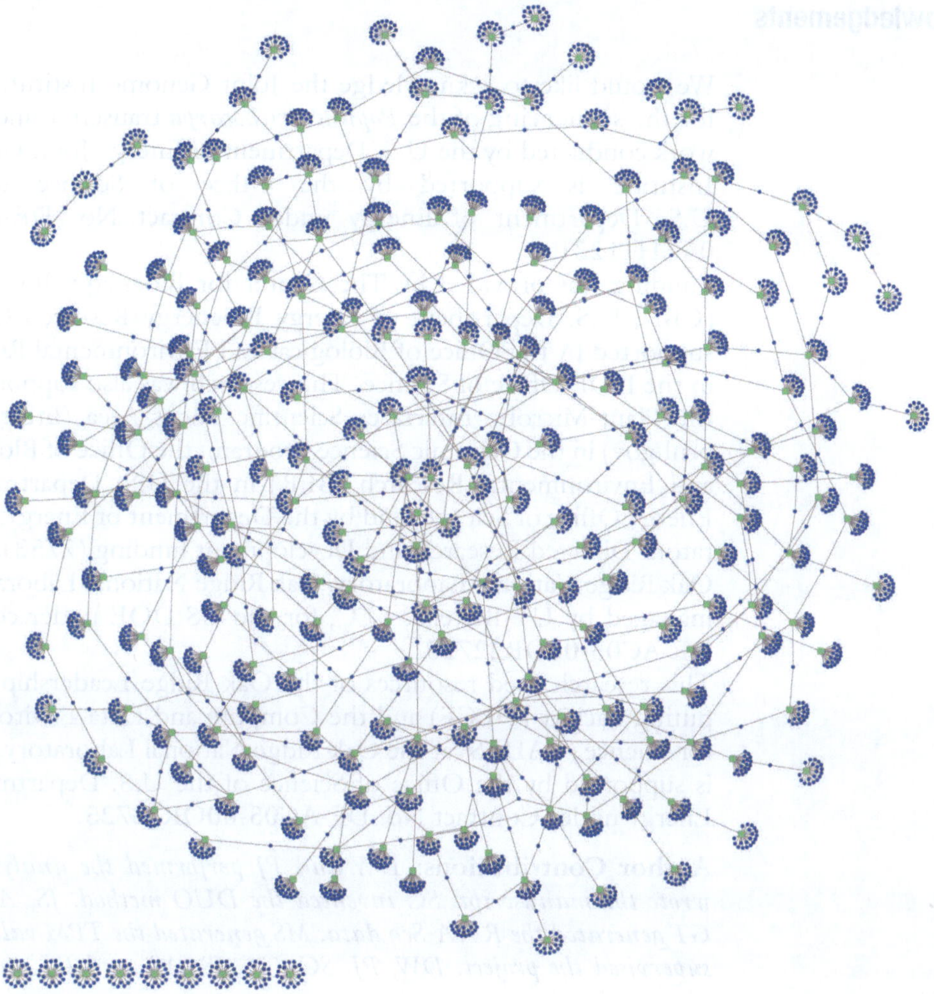

Fig. 7 Random forest network. Network visualization of the top 20 important variables in an all pairs comparison of samples, grouped by replicates. The underlying data is from the *P. trichocarpa* gene atlas gene expression data set. Round purple nodes indicate genes, while square green nodes indicate the group comparisons that was performed. The network is mostly connected by means of one or two shared genes between certain groups comparisons. Though, as expected there are some disjoint groupings as well

Fig. 6 (continued) comparison. The color of the edge indicates whether or not the gene was over (red) or under expressed (blue) in the comparison. The over/under association is determined relative to the sample name that appears first in the comparison node label. The color intensity of the edge correlates with the absolute log fold change of expression. The box depicts the genes connected to the indicated leaf comparison nodes, here two genes are highlighted. **(B)** Lines plot for the two highlighted genes under the three conditions described by the two comparison nodes. The *x*-axis indicates replication, *y*-axis the TPM value and each line represents a particular sample. Here we see that the algorithm reveals both dramatic differences between conditions as well as more subtle differences

Acknowledgements

We would like to acknowledge the Joint Genome Institute (JGI) for the sequencing of the *Populus trichocarpa* transcriptomes. The work conducted by the U.S. Department of Energy Joint Genome Institute is supported by the Office of Science of the U.S. Department of Energy under Contract No. DE-AC02-05CH11231.

Funding was provided by The Center for Bioenergy Innovation (CBI), U.S. Department of Energy Bioenergy Research Centers supported by the Office of Biological and Environmental Research in the DOE Office of Science. This research was also supported by the Plant-Microbe Interfaces Scientific Focus Area (http://pmi.ornl.gov) in the Genomic Science Program, the Office of Biological and Environmental Research (BER) in the U.S. Department of Energy Office of Science, and by the Department of Energy, Laboratory Directed Research and Development funding (7758), at the Oak Ridge National Laboratory. Oak Ridge National Laboratory is managed by UT-Battelle, LLC, for the US DOE under contract DE-AC05-00OR22725.

This research used resources of the Oak Ridge Leadership Computing Facility (OLCF) and the Compute and Data Environment for Science (CADES) at the Oak Ridge National Laboratory, which is supported by the Office of Science of the U.S. Department of Energy under Contract No. DE-AC05-00OR22725.

Author Contributions: *DW and PJ performed the analysis and wrote the manuscript. SC invented the DUO method. JS, AS, and GT generated the RNA-Seq data. MS generated the TPM values. DJ supervised the project. DW, PJ, SC, GT, JS, AS, and DJ edited the manuscript.*

References

1. Barabasi A-L, Oltvai ZN (2004) Network biology: understanding the cell's functional organization. Nat Rev Genet 5(2):101
2. Balakrishnan R, Ranganathan K (2012) A textbook of graph theory. Springer Science & Business Media, New York
3. Zheng Q, Wang X-J (2008) Goeast: a web-based software toolkit for gene ontology enrichment analysis. Nucleic Acids Res 36 (suppl_2):W358–W363
4. Langfelder P, Horvath S (2008) WGCNA: an r package for weighted correlation network analysis. BMC Bioinf 9(1):559
5. Movahedi S, Van Bel M, Heyndrickx KS, Vandepoele K (2012) Comparative co-expression analysis in plant biology. Plant Cell Environ 35 (10):1787–1798
6. Van Dongen SM (2001) Graph clustering by flow simulation. Doctoral dissertation, PhD thesis, University of Utrecht
7. Van Dongen S (2008) Graph clustering via a discrete uncoupling process. SIAM J Matrix Anal Appl 30(1):121–141
8. Climer S et al (2020) Discovery of synchronized gene expression modules using a vector-based correlation coefficient. bioRxiv doi: https://doi.org/10.1101/2020.01.28.923730
9. Climer S, Yang W, Fuentes L, Dávila-Román VG, Gu CC (2014) A custom correlation coefficient (CCC) approach for fast identification

of multi-SNP association patterns in genome-wide SNPs data. Genet Epidemiol 38 (7):610–621

10. Climer S, Templeton AR, Zhang W (2014) Allele-specific network reveals combinatorial interaction that transcends small effects in psoriasis GWAS. PLoS Comput Biol 10(9): e1003766
11. Dembélé D, Kastner P (2014) Fold change rank ordering statistics: a new method for detecting differentially expressed genes. BMC Bioinf 15(1):14
12. Breiman L (2001) Random forests. Mach Learn 45(1):5–32
13. Dietterich TG (2000) An experimental comparison of three methods for constructing ensembles of decision trees: bagging, boosting, and randomization. Mach Learn 40 (2):139–157
14. Banerjee S, Pedersen T (2003) The design, implementation, and use of the Ngram statistics package. In: Gelbukh A (ed) Computational linguistics and intelligent text processing. CICLing 2003. Lecture notes in computer science, vol 2588. Springer, New York, pp 370–381
15. R Core Team (2015) R: a language and environment for statistical computing. R Foundation for Statistical Computing, Vienna
16. RStudio Team (2016) RStudio: integrated development environment for R. RStudio, Inc., Boston, MA
17. Dowle M, Srinivasan A (2017) data.table: Extension of 'data.frame'. R package version 1.10.4
18. Wickham H (2009) ggplot2: elegant graphics for data analysis. Springer, New York
19. Arnold JB (2017) ggthemes: extra themes, scales and geoms for 'ggplot2'. R package version 3.4.0
20. Chen WC, Ostrouchov G, Schmidt D, Patel P, Yu H (2012) pbdMPI: programming with big data–interface to MPI. R Package, http://cran.r-project.org/package=pbdMPI
21. Liaw A, Wiener M (2002) Classification and regression by randomforest. R News 2 (3):18–22
22. Wickham H (2007) Reshaping data with the reshape package. Journal of statistical software 21.12:1–20.
23. Shannon P, Markiel A, Ozier O, Baliga NS, Wang JT, Ramage D, Amin N, Schwikowski B, Ideker T (2003) Cytoscape: a software environment for integrated models of biomolecular interaction networks. Genome Res 13(11):2498–2504
24. Jiang H, Lei R, Ding S-W, Zhu S (2014) Skewer: a fast and accurate adapter trimmer for next-generation sequencing paired end reads. BMC Bioinf 15(1):182
25. Dobin A, Davis CA, Schlesinger F, Drenkow J, Zaleski C, Jha S, Batut P, Chaisson M, Gingeras TR (2013) Star: ultrafast universal RNA-seq aligner. Bioinformatics 29(1):15–21
26. Benjamini Y, Hochberg Y (1995) Controlling the false discovery rate: a practical and powerful approach to multiple testing. J R Stat Soc Ser B Methodol 57:289–300
27. Law CW, Chen Y, Shi W, Smyth GK (2014) Voom: precision weights unlock linear model analysis tools for RNA-seq read counts. Genome Biol 15(2):R29
28. Tuskan GA, Difazio S, Jansson S, Bohlmann J, Grigoriev I, Hellsten U, Putnam N, Ralph S, Rombauts S, Salamov A et al (2006) The genome of black cottonwood, Populus trichocarpa (torr. & gray). Science 313 (5793):1596–1604
29. Goodstein DM, Shu S, Howson R, Neupane R, Hayes RD, Fazo J, Mitros T, Dirks W, Hellsten U, Putnam N et al (2012) Phytozome: a comparative platform for green plant genomics. Nucleic Acids Res 40(D1): D1178–D1186
30. Weighill DA, Jacobson DA (2015) 3-Way networks: application of hypergraphs for modelling increased complexity in comparative genomics. PLoS Comput Biol 11(3): e1004079
31. Weighill DA, Jacobson D (2016) Network metamodeling: effect of correlation metric choice on phylogenomic and transcriptomic network topology. In: Nookaew I. (eds) Network Biology. Advances in Biochemical Engineering/Biotechnology, vol 160. Springer, Cham
32. Fujita A, Sato JR, Demasi MA, Sogayar MC, Ferreira CE, Miyano S (2009) Comparing Pearson, Spearman and Hoeffding's D measure for gene expression association analysis. J Bioinform Comput Biol 7(04):663–684
33. Bloom SA (1981) Similarity indices in community studies: potential pitfalls. Mar Ecol Prog Ser 5:125–128
34. Chen Y, McCarthy D edgeR: differential expression analysis of digital gene expression data user's guide. Available online: http:// www. bioconductor. org/packages/release/ bioc/vignettes/edgeR/inst/doc/edgeRUsersGuide.pdf

Chapter 16

Connecting Microbial Genotype with Phenotype in the Omics Era

Yongfu Yang, Mengyu Qiu, Qing Yang, Yu Wang, Hui Wei, and Shihui Yang

Abstract

Although rational design-based metabolic engineering has been applied widely to obtain promising microbial biocatalysts, conventional strategies such as adaptive laboratory evolution (ALE) and mutagenesis are still efficient approaches to improve microorganisms for exceptional features such as a broad spectrum of substrate utilization, robustness of cell growth, as well as high titer, yield, and productivity of the target products. In this chapter, we describe the procedure to generate mutant strains with desired phenotypes using ALE and a new mutagenesis approach of Atmosphere and Room Temperature Plasma (ARTP). In addition, we discuss the methodology to combine next-generation sequencing (NGS)-based genome-resequencing and RNA-Seq transcriptomics approaches to characterize the mutant strains and connect the phenotypes with their corresponding genotypic changes.

Key words Adaptive laboratory evolution (ALE), Atmosphere and Room Temperature Plasma (ARTP) mutagenesis, Next-generation sequencing (NGS), RNA-Seq, CLC Genomics Workbench, JMP Genomics

1 Introduction

Microorganisms play critical roles in various areas such as bioenergy, bioremediation, biochemical, and pharmaceutical productions. Microbial strains with exceptional features of broad substrate utilization, robustness against various inhibitors, as well as high titer, yield, and productivity of the target products are the key for cost-efficient biomanufacturing. With recent, rapid biotechnology advances on "genome reading" using next-generation sequencing (NGS) techniques as well as "genome writing" using novel genome editing tools such as CRISPR/Cas9 and synthetic biology strategies, rational design-based metabolic engineering has been applied extensively to obtain promising industrial phenotypic characteristics directly by incorporating heterologous gene(s) or pathway

Michael E. Himmel and Yannick J. Bomble (eds.), *Metabolic Pathway Engineering*, Methods in Molecular Biology, vol. 2096, https://doi.org/10.1007/978-1-0716-0195-2_16, © Springer Science+Business Media, LLC, part of Springer Nature 2020

(s) into the chassis strains for novel microbial biocatalyst development.

However, our current understanding on the biological parts and especially their regulation in the background of complicated cellular metabolism and regulatory network is still inadequate to build a desired synthetic microorganism from scratch for the expected phenotype. For example, sometimes it is more effective to target regulatory factors rather than key enzyme genes to increase the productivity of target bioproducts [1, 2]. Therefore, conventional strategies using mutagenesis and natural selection approaches such as adaptive laboratory evolution (ALE) before or after metabolic engineering are still the effective approaches for strain improvement with a combination of different excellent genotypic changes for industrial applications.

ALE strategy, developed about a 100 years ago, is a powerful method to improve certain features of common industrial strains without a prior knowledge of any underlying genetic mechanisms, as long as the desired traits can be coupled with growth [3]. During the ALE process, a microorganism is cultivated under the defined selective pressure conditions for prolonged periods of time from weeks to years, allowing the selection of improved phenotypes. To resolve the limitation of low spontaneous mutation rate using ALE natural selection [4], various mutagenesis approaches, such as chemical mutagenesis, transposon mutagenesis, genome shuffling, or error-prone PCR, have been further developed and used for microbial strain improvement.

Recently, a new method termed as Atmosphere and Room Temperature Plasma (ARTP) has been developed by Tmaxtree Biotechnology Co. Ltd. (Wuxi, China) to generate mutant library efficiently and safely. ARTP mutagenesis system is designed with the plasma generator as the core component and helium (with purity of 99.99% or better) as the working gas. When helium gas flowing through two energized electrodes, high energy electrons and the surrounding neutral particles undergo energy exchange through elastic collision and inelastic collision, making helium decomposed, stimulated, or ionized. With the voltage increasing, the gas is discharged continually and broken down to form a plasma with a certain degree of ionization. With the effect of airflow, the plasma jet is ejected at a certain speed and reacts with the air components, which contain nitrogen, hydroxyl, and other active particles [5]. These chemically reactive species are still abundant even at a low temperature close to room temperature and suitable for the treatment of biomolecules or cells. The helium plasma jet has been conformed to change the DNA sequences significantly, thus overcoming the limitations of traditional methods, such as high and uncontrollable gas temperatures, low efficiency, time consuming, and operation difficulty [5, 6]. Recently, using *umu* test strategy and flow-cytometric analysis, ARTP has been confirmed that it produced greater DNA damage to individual living cells with higher mutation rates than conventional chemical and physics mutagenesis [6]. Moreover, ARTP has been experimentally demonstrated as a powerful mutagenesis tool to improve the desired industrial characteristics of various microbial cell factories such as *E. coli* and *Bacillus subtilis* [7–15].

Prior to the development and prevailing applications of NGS techniques, it was nearly impossible to correlate phenotypic changes generated from natural selection and mutagenesis with genotypic changes since different types of mutations can be generated during the mutagenesis, and mutation sites are nearly unpredictable. For example, the genetic changes of an acetate-tolerant mutant of *Zymomonas mobilis* generated from chemical mutagenesis in 1998 [16] were not revealed till 2010 through microarray-based comparative genome-sequencing and transcriptomic study [17]. With the rapid NGS techniques development and decreasing sequencing cost, NGS-based genome-resequencing has been used dominantly in microbial mutant characterization. For example, genotypic changes associated with the hydrolysate tolerance and the improved arabinose utilization in *Z. mobilis* strains were reported [18, 19].

Although the NGS-based genome-resequencing technology could reveal the genetic changes associated with mutants generated from different mutagenesis approaches, further work using systems biology approaches is still needed to connect these genetic changes with phenotypic changes. For example, although our genome-resequencing work has identified the genetic changes associated with the acetate tolerant mutant AcR [17], it was still not clear whether large genetic changes such as deletion/insertion/inversion or small genetic changes of SNPs played a major role in acetate tolerance. Furthermore, even we confirmed that the large deletion affects the coding region of gene ZMO0117 and the promoter region of a Na^+/H^+ antiporter coding gene ZMO0119 (*nhaA*), it was also not clear whether the deletion of ZMO0117 or the truncated ZMO0119 promoter contributes to the acetate tolerance phenotype [17]. Genetics study was carried out and confirmed the association of *nhaA* upregulation with the improved acetate tolerance phenotype in acetate mutant AcR at the guidance of the transcriptomic result that the *nhaA* gene was constitutively upregulated about 16 folds in the mutant background compared with parental strain [17]. Interestingly, another acetate-tolerant mutant ZMA-167 reported recently possessed a similar genotype change closed to the AcR mutant [20], which further confirmed the success and the necessity to combine the genome-resequencing with other systems biology approaches to connect the phenotypic changes with the genotypic alterations.

In this chapter, we describe the procedure to generate mutant strains with the desired phenotype and the subsequent approaches to characterize the mutant strains to connect the phenotype with their corresponding genotypic changes using NGS-based genome-resequencing and RNA-Seq transcriptomics studies, as well as software for NGS result analysis.

2 Materials

All buffers and media are prepared using distilled and deionized water with a resistivity of 18.25 MΩ-cm at 25 °C and sterilized either by autoclaving or filtration through a 0.22-μm membrane, which are stored at room temperature unless indicated otherwise. Chemicals are analytical grade. Waste disposal regulations are followed for handling waste materials.

2.1 Strains and Media

1. Strain: wild-type *Zymomonas mobilis* ATCC31821 (ZM4).
2. RMG (rich media with glucose) medium:50 g/L glucose, 10 g/L yeast extract, 2 g/L KH_2PO_4, pH 5.8.

2.2 General Chemicals and Equipment

1. ARTP-M mutation breeding system.
2. PCR and real-time PCR machines.
3. High-pressure steam autoclave.
4. Centrifuge and refrigerated centrifuge.
5. Shaker.
6. Incubator.
7. NanoDrop Spectrophotometer (Thermo Fisher Scientific Inc., MA, USA).
8. Five water baths for temperatures of 70 °C, 65 °C, 60 °C, 55 °C, 37 °C.
9. Sterilized 1.5 mL and 2 mL microcentrifuge tubes and nuclease-free tubes.
10. Inoculating loop.
11. Inoculation spreader.
12. Stainless steel slide.
13. Forceps.
14. Culture dish.
15. Vortex mixer.
16. Glass beads.
17. 0.5 mM EDTA (pH = 8.0).
18. Isopropyl alcohol.
19. Chloroform.
20. RNase-free water.

21. Glycerol.
22. 0.85% NaCl.
23. 75% ethanol and 75% RNase-free ethanol.
24. RNase-free ethanol.
25. DNase I buffer (NEB#B0303S).
26. DNase Enzyme (NEB#M0303S).
27. RNaseOUTTM Recombinant ribonuclease inhibitor (Invitrogen #10777).
28. dNTPs,
29. SuperScript III Reverse Transcriptase (Invitrogen #18080044).
30. 0.1 M DTT.
31. RNase H (NEB #R0297S).
32. iQ SYBR Green Super mix.
    (a) 1000 × 50 μL reactions (Bio-Rad cat. 170-8884).
    (b) 500 × 50 μL reactions (Bio-Rad cat. 170-8882).
33. RT-PCR plate (Bio-Rad, cat. 223-9441).
34. 25 mM $MgCl_2$
35. Taq polymerase (NEB#M0273S).
36. iCycler iQ™.
37. Optical Quality sealing tape (Bio-Rad, cat. 223-9444).
38. SYBR Green (Molecular Probes, cat. S-7567).

2.3 Related Kits

1. TIANGEN® TIANamp Bacteria DNA Kit.
2. TRIzol® Reagent.
3. Qiagen RNeasy Mini Kit.

3 Methods

Carry out all procedures at room temperature unless otherwise specified.

3.1 Mutagenesis Through ARTP and Adaptive Laboratory Evolution

3.1.1 ARTP Mutagenesis and ALE

Due to the flexible operation, low cost, high mutation rate, and wide range of applications, mutagenesis techniques are still widely used for obtaining microbial strains with excellent industrial characteristics. ARTP mutation breeding technology has the following advantages: (1) simple equipment structure and low operation cost; (2) high mutation rate with multiple mutagenesis mechanisms; (3) easy manipulation, and high safety. Therefore, we use the ARTP-M mutant system as an example to describe a simple mutagenesis operation process. ALE is another powerful method to

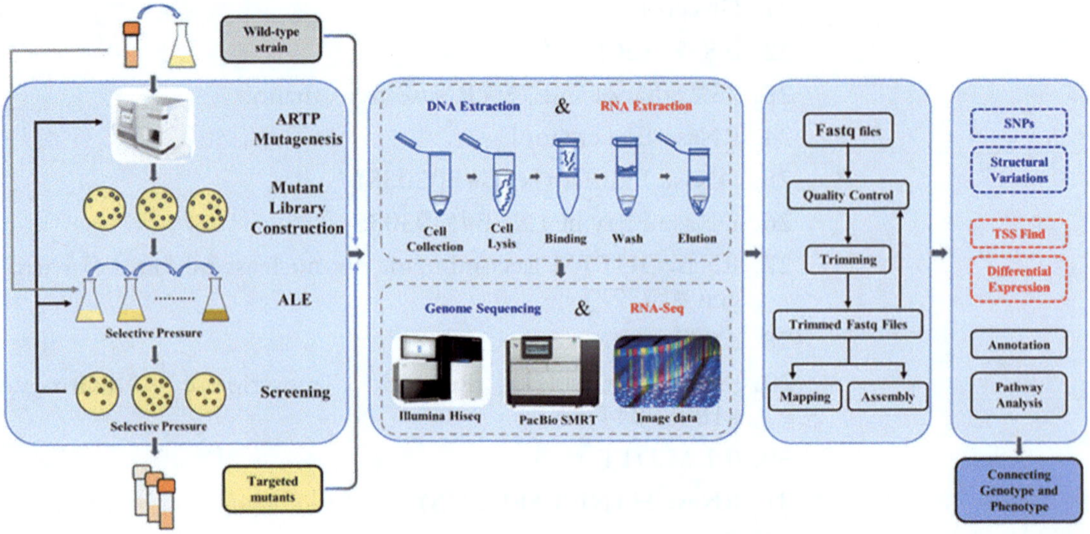

Fig. 1 The pipeline to connect phenotypic change(s) with genotype including the process to obtain mutant strains using ALE and ARTP mutagenesis, to characterize mutant strains using NGS-based genome-resequencing and RNA-Seq, as well as to analyze NGS result for phenotype–genotype correlation

improve industrial features without prior knowledge of any underlying genetic mechanisms. Either ARTP or ALE alone (or their combination) can be used to select strains with the desired phenotype(s) (*see* Fig. 1 for details).

1. Sterilize the related experimental materials and reagents by autoclaving or filtration.
2. Activate bacterial strain: Take wild-type *Z. mobilis* ZM4 cells from frozen glycerol stock and activate it by streaking on RMG plate, pick up single colony by inoculating loop from plate into RMG medium and cultivate it to log phase in a shaker at 30 °C, 100 rpm.
3. Prepare bacterial suspension: centrifuge 1 mL log-phase culture sample with OD_{600nm} value at 0.6–0.8, wash cell pellets two to three times using 0.85% NaCl, 5% glycerol to resuspend the culture.
4. Power on ARTP-M mutation system and open gas path, check the running status, wipe the cab by 75% ethanol and treat it with ultraviolet irradiation.
5. Add 10 μL bacterial suspensions to stainless steel slide surface and spread, move the slide to the cab by sterile culture dish and forceps.
6. Set up power, time, gas volume, and run the instrument, followed by putting each processed slide to 2 mL microcentrifuge tubes with 1 mL medium, respectively (*see* Subheading

3.1.2 below for details on choosing optimal ARTP mutagenesis parameters).

7. Drain the residual gas, close the gas path, wipe the cab by 75% ethanol and treat it with ultraviolet irradiation, and power off ARTP-M mutation system.
8. Put 2 mL microcentrifuge tubes on the vortex mixer for 1 min and wash the adhesion bacteria to the RMG broth.
9. Dilute the culture suspension with a ten-fold dilution ratio to make 10^{-2}, 10^{-3}, and 10^{-4} dilutions, and transfer and spread 100 μL of the diluted samples on RMG plates by using glass beads in triplicate for each sample.
10. Cultivate the microorganism and then count the number of colonies grown on each plate. Calculate the fatality rate using the following formula: Fatality rate (%) = [1 − (the number of colonies for ARTP-treated samples/the number of colonies of blank control sample)] × 100%.
11. For ALE, microorganism is cultivated, and an aliquot of log-phase culture sample is serially transferred into fresh media under selective pressure conditions continuously until the desired improved phenotype appears, which could be in the range of weeks to years.

3.1.2 ARTP Optimal Parameters Determination and Mutant Library Construction

1. The variable parameters of ARTP are power, time, and gas volume. The optimal fatality rate is about 90%. In the process of ARTP mutagenesis for ZM4, the gas volume is set at 10 standard liter per minute (SLM), which can be adjusted at the range of 8–12 SLM for other microorganisms.
2. Other parameters are then optimized based on recommended parameters for different microorganisms (Table 1). For example, we fix the treatment time of 15 s and gas volume of 10 SLM, and then investigate the fatality rate using different treatment powers of 90, 100, 110, or 120 W.

Table 1
Recommended sample processing time according to ARTP-M manual with other parameters of 120 w, 10 SLM, and 2 mm (distance between stainless steel slide and the mutagen)

Microorganism		Suggested test time (s)
Prokaryotes	Bacteria	15, 30, 45, 60, 90, 120
	Actinomycetes	30, 60, 90, 120, 150, 180
Eukaryotes	Fungi	60, 120, 180, 240, 300
	Yeast	30, 60, 90, 120, 150, 180, 240
	Microalgae	5, 10, 15, 20, 30, 40, 50, 60, 90, 150

3. After the treatment power is determined (e.g., 90 W for *Z. mobilis*), we then investigate a series of different treatment times (0, 5, 10, 15, 25, 35, 45, or 60 s) and determine the optimal treatment time (e.g., 15 s for *Z. mobilis*).
4. Taking preliminary results of these experiments, we choose 90 W and 15 s as the mutagenesis condition to treat the cell suspension for mutant library construction.
5. Put 1 mL of strain suspension into 2-mL micro-centrifuge tubes and vortex for 1 min.
6. Transfer above suspension onto RMG plates (100 μL per plate), and then cultivate at 30 °C incubator until obvious colonies appear. Add about 1 mL 30% glycerol to each plate and scrape off colonies using inoculation spreader. Mix the glycerol solution obtained from above plates, divide into aliquots (1.5 mL per vial), and store at −80 °C freezer.
7. For the selection of mutant strains against different stressors, such as low pH, selective media can be used. For example, to screen the mutant strain resistant to acid tolerance, the above mutant library can be spread on pH 3.6 RMG plates to select mutant strains with enhanced acid tolerance.

3.2 Bacterial Genome DNA Extraction

The method is based on the technical manual of TIANGEN® TIANamp Bacteria DNA Kit (Beijing, China).

1. Take freeze glycerol stock from −80 °C refrigerator, activate strain by streaking lines on RMG plate and cultivate at 30 °C incubator for about 2 days.
2. Pick up single colony from plate into RMG medium and culture at 30 °C, 100 rpm on a shaker, until the OD_{600nm} value reaches about 1.0.
3. Pellet 2–4 mL of bacterial culture by centrifugation 1 min at top speed (13,500 × *g*) in a microcentrifuge. Discard the supernatant and remove any excess media.
4. Add 200 μL of GA buffer to completely resuspend the cell pellet by pipetting or vortexing (*see* **Notes 1** and **2**).
5. Add 20 μL of Proteinase K and mix by inverting the tube five times.
6. Add 220 μL of GB buffer to the sample, vortex for 15 s, and incubate it in heating block at 70 °C for 10 min to yield a homogeneous solution. The solution should become clear. Briefly centrifuge the centrifuge tube to remove residual solution from the inside of the tube lid (*see* **Note 3**).
7. Add 220 μL of 100% ethanol and mix thoroughly by shaking for 15 s; floccule should appear in the solution and centrifugate briefly to remove residual solution on the wall of tube.

8. Insert the spin-column CB3 into a collection tube, carefully transfer the full volume of floccule-containing solution at **step** 7 directly to spin-column, and centrifuge at top speed for 30 s. Discard the filtrate.
9. Add 500 μL of GD buffer (add 100% ethanol into the GD before use) to the spin-column, and centrifuge at top speed for 30 s. Discard the filtrate.
10. Add 600 μL of PW buffer (add 100% ethanol before use) to the spin-column, and centrifuge at top speed for 30 s. Discard the filtrate. Repeat this step one more time.
11. Centrifuge at top speed for additional 2 min and discard the filtrate. Incubate the spin-column in room temperature for 2–3 min to remove residual trace of PW (*see* **Note 4**).
12. Transfer the spin-column into a new microcentrifuge tube and add 50–200 μL of TE or H_2O into the column and wait for 2–5 min.
13. Centrifuge at top speed for 1 min to elute the genome DNA (*see* **Note 5**).
14. Quantify the DNA using NanoDrop Spectrophotometer and check the DNA quality by 1% agarose gel electrophoresis with a loading of about 50 ng genomic DNA per vial.
15. If the DNA has the 260/280 and 260/230 ratios greater than 1.7 and no smear in the gel, store the genomic DNA at −20 °C until use.

3.3 Bacterial RNA Extraction

The method is based on the technical manual of TRIzol® Reagent and Qiagen RNeasyMini Kit. Always use the appropriate precaution to avoid RNase contamination when preparing and handling RNA.

1. Use 1 mL of the TRIzol Reagent (in the fridge) per 1 × 10^7 bacterial cells.
2. Pipette and vortex to lyse cell (*see* **Note 6**).
3. Incubate the homogenized samples for 5 min at room temperature.
4. Add 0.2 mL of chloroform, cap and shake for 15 s and incubate them at room temperature for 2–3 min.
5. Centrifuge the samples at 6000 × *g* for 15 min at 4 °C.
6. Transfer the aqueous RNA phase to a fresh tube (*see* **Notes** 7 and **8**).
7. Add 0.5 mL of isopropyl alcohol to precipitate RNA (*see* **Note 9**).
8. Incubate samples at room temperature for 10 min, centrifuge at 6000 × *g* for 10 min at 4 °C (forms a gel-like pellet on the side and bottom of the tube) and remove the supernatant.

9. Wash the RNA pellet once with 1 mL of 75% RNase-free ethanol and mix by vortexing.
10. Centrifuge at 2400 × *g* for 5 min at 4 °C.
11. Dry the RNA pellet briefly in air, approximately 10 min.
12. Dissolve RNA in 50 μL of RNase-free water by passing the solution a few times through a pipette tip.
13. Incubate for 10 min at 55–60 °C.
14. Set heat block to 65 °C.

    Add into a tube:

    10 μL DNase I buffer.

    2 μL DNase enzyme.

    50 μL RNA sample.

    38 μL H_2O to bring to 100 μL total volume.

    Mix gently and incubate at 37 °C for 20–30 min.
15. Add 1 μL of 0.5 mM EDTA (pH 8.0), incubate at room temperature 1 min and incubate at 65 °C for 10 min.

3.3.1 Proceed with Qiagen RNeasy Kit

1. Add 350 μL of Buffer RLT and mix.
2. Add 250 μL RNase-free ethanol (96–100%) and mix thoroughly by pipetting.
3. Apply the sample (about 700 μL) to a RNeasy mini column.
4. Centrifuge for 15 s at 13,500 × *g* and transfer to a new 2 mL collection tube.
5. Add 500 μL Buffer RPE and centrifuge for 15 s at 13,500 × *g* and discard the flow-through completely.
6. Add another 500 μL of Buffer RPE and centrifuge at 13,500 × *g* for 2 min.
7. Place the RNeasy column in a new 2-mL collection tube and centrifuge at full speed for 1 min.
8. Place the RNeasy column in a new tube.
9. Pipette 30 μL of RNase free water directly onto the RNeasy silica-gel membrane.
10. Centrifuge for 1 min at 13,500 × *g*.
11. Use NanoDrop Spectrophotometer to confirm the concentration of RNA elute before discarding the RNeasy column.
12. Repeat elution if concentration of RNA is too low (*see* **Note 10**).

3.4 Rt-Pcr

The method is based on Qiagen RNeasy kit.

3.4.1 Prepare Gene-Specific Primer Sets

1. Select genes or sequences for real-time quantitative PCR.
2. Use the web-based primer design program 'Primer3' to design primer sets specifying a PCR product of 100 bp, 60% > GC > 40%, $T_m \sim 55$ °C.
3. Synthesize the primers.
4. Prepare a mixture of forward and reverse primers at a final concentration of 5 pmol/μL each.

3.4.2 Prepare Samples for Standard Curves

1. Once obtaining the primer sets, amplify the corresponding fragments using common PCR machine to make sure all primers are working and producing only one amplicon of interest.
2. Purify the amplified product using Qiagen spin columns.
3. Use NanoDrop Spectrophotometer to determine DNA concentration of the purified PCR product.
4. Prepare standards (10^8 copies of known DNA molecule per μL as starting concentration) and a dilution series with known DNA molecule copy number for a standard curve (10^7 to 10^1 copies per μL works well).
5. Use the DNA concentration at A_{260} to calculate copy number of the DNA molecules.

$$\text{Copies } (\mu\text{L}) = A_{260} \times 75.8 \text{ (size of product in bp)} \times 6 \times 10^{11} \text{ copies } (\mu\text{L})$$

6. The dynamic range of standard preps for iCycler is from 2000 to 20,000,000 copies, which can be extrapolated into a few copies.
7. Store samples at −20 °C until ready to use.

3.4.3 cDNA Preparation

1. Prepare cDNA from total cellular RNA by reverse transcription.

Total RNA	200 ng
Random primers	100 ng (Invitrogen 50 ng/μL use 2.0 μL)
dNTPs	10 mM
ddH$_2$O	μL (bring to final volume of 20 μL)
Total	20 μL

2. Incubate at 65 °C for 5 min.
3. Transfer to ice to cool.

4. Add the following to each tube.

5× First-stand Buffer	4 μL
0.1 M DTT	2 μL
RNaseOUTTM recombinant ribonuclease inhibitor (Invitrogen #10777)	1 μL
SuperScript III reverse transcriptase (Invitrogen #18080044)	1 μL
ddH_2O	μL (bring to final volume of 20 μL)
Total	20 μL

5. Incubate at 55 °C for 1 h.
6. Incubate at 70 °C for 15 min.
7. Transfer to ice to cool.
8. Treated with RNase H (NEB #R0297S) for 20 min at 37 °C.
9. Store at −20 °C.
10. Dilute a portion of cDNA as a working stock to ~75 ng/μL.

3.4.4 Real-Time PCR (Using SYBR Green Super Mix)

1. Create and save the PCR program in the 'Protocol Workshop'. In the 'Edit Protocol' window, the thermal parameters are set as following (*see* **Note 11**):

Step 1:	95 °C	3 min 15 s	1 cycle
Step 2:	95 °C 55 °C 72 °C	15 s 30 s 30 s	40 cycles
Step 3:	95 °C	60 s	
Step 4:	55 °C	10 s	80 cycles for melting curve
Step 5:	4 °C		Hold forever

2. Create and save the plate setup in 'Protocol Workshop'. 'Edit Plate Setup' allows to specify the positions of test samples and standard samples. Mark wells with specific fluorophores such as FAM-490 for SYBR green I. Fill in the quantity and units (weight or copy number) for each individual standard well.
3. Add template and standard to plate.

4. Prepare master mix as follows and distribute 29 μL to each well in a thin wall PCR plate (Bio-Rad, cat. 223-9441) (*see* **Note 12**).

Super mix	Per rxn	Per plate (120 rxns)
iQ SYBR green super mix	15 μL	1800 μL
Primer 1	1 μL (5 pmol/μL)	120 μL
Primer 2	1 μL (5 pmol/μL)	120 μL
Sterile H_2O	12 μL	1440 μL
DNA template	1 μL	Add individually

SYBR GREEN I	Per rxn	Per plate (120 rxns)
DNA template	1.0 μL	Add individually
Sterile H_2O	19.6 μL	2352 μL
10 × PCR buffer	3.0 μL	360 μL
25 mM $MgCl_2$	1.8 μL	216 μL
10 mM dNTP's	0.6 μL	72 μL
Primer 1	1.0 μL (5 pmol/μL)	120 μL
Primer 2	1.0 μL (5 pmol/μL)	120 μL
Taq polymerase	1.0 μL	120 μL
1 × SYBR green	1.0 μL	120 μL

5. Once master mix is added, place on ice and prevent exposure to light by wrapping it with aluminum foil.
6. Prepare external well factor plate. If using super mix, the well factors are in the super mix so just put sample plate in, and no special well factor plate needs to be created. If SYBR green I is used, dilute external well factor plate solution by a 1:10 in PCR buffer and load 30 μL into each well of a thin wall PCR plate. Cover with the optical quality sealing tape. Also prevent light by wrapping it with aluminum foil.
7. Use 'Imaging service' to confirm that 1× well factor solution gives a strong but not saturated image.
8. Check PCR program and plate setup in the 'View Plate Setup' tab before running.
9. Check to make sure iCycler, camera, and the filter selected are properly switched on or in the appropriate position.
10. Place the well factor plate with the external well factor plate solution or super mix plate into iCycler and select correct PCR program and plate setup.

11. Click 'Run'.
12. In the 'Run Prep' tab, confirm once again the desired protocol and plate setup file. Enter the reaction volume (30 μL). Indicate the type of protocol (PCR Quantification/Melt curve) and the Well Factor Source, then click 'Begin Run'.
13. If using a SYBR green I protocol in about 5 min, the iCycler will go into Pause mode. During the pause period, you should remove the well factor plate (which can be transferred to −80 °C freezer for reuse later), replace it with the experimental PCR plate or just continue run if using super mix (*see* **Note 13**).
14. Click 'Continue Running Protocol'.
15. After data collection on the PCR reaction plate begins, the PCR Amp Cycle plot will be displayed, and the software will open the data analysis module.
16. Calculate copy number of the DNA molecules according to standard curves.

3.5 Next-Generation Sequencing (NGS)-Based Genome-Sequencing and RNA-Seq

Genomic DNA or total RNA with good quality based on agarose gel electrophoresis and bioanalyzer analyses are sent to NGS-service provider such as Genewiz for library construction and sequencing; the fastq files generated from NGS will be evaluated for their quality before bioinformatics and statistical analyses to interpret data and correlate phenotype with genotype. Two different approaches can be used: commercial software such as CLC Genomics Workbench or open-source software.

4 Bioinformatics and Statistical Analyses

1. Quality control (QC) is crucial to avoid "garbage in and garbage out." The raw FASTQ data are first imported into the CLC Genomics Workbench (Qiagen, CA, USA) with the default setting. Subsequently, selecting the "Create Sequencing QC Report" under the "NGS Core Tools" submenu in the "Toolbox" drop-down list to create the QC report. Alternatively, the standalone open-source software FastQC (http://www.bioinformatics.babraham.ac.uk/projects/fastqc) can be used to check the reads quality.
2. Low quality reads based on QC report, which are usually in the beginning or at the end of each reads, are trimmed through the "Trim Sequences" function in "NGS Core Tools" submenu. Quality control should be run again to review the quality after trimmed. Similarly, the standalone software Trimmomatic [21] can also be used to pre-process NGS-based data (bath genome-

sequencing and RNA-Seq) to remove contaminants, adaptors, low-quality sequences, and other artifacts.

3. The trimmed reads with good quality are then used for genome mapping as well as de novo assembly using the corresponding modules which located at "NGS Core Tools" or "RNA-Seq Analysis" submenu in CLC Genomics Workbench software. Similarly, alternative strategy using open-source software can also be used in this step. For example, Bowtie2 [22] can be used to map trimmed reads to reference genome with the SAM format output file followed by SAMtools [23] to compress the SAM files to BAM format creating files with ordered and indexed sequences. Software SOAP denovo2 (http://soap.genomics.org.cn/) [24] can be used for de novo assembly, and GapCloser v1.12 software can then be used to fill the gaps and correct the bases on the assembly results. For microorganisms without sequenced reference genome, Trinity [25] can be used for de novo assembly.
4. If reads are from RNA-Seq, they are used for RNA-Seq analysis, which are located at "RNA-Seq Analysis" submenu in CLC Genome Workbench, and for generating the gene expression level such as RPKM (Reads Per Kilobase per Million mapper reads) value for each gene in the genome. Open-source software such as BWA, Bowtie2, TopHat [26], and Hisat [27] can be used for mapping, and StringTie [28] can be used for transcript assembly with reference genome.
5. For differential gene expression analysis, JMP Genomics from SAS Inc. (NC, USA) is used. Data generated from above bioinformatics analysis such as the RPKM values will be imported into JMP Genomics. Data quality is then examined and further normalized before statistical modeling such as ANOVA is applied. Alternatively, open-source software such as the R package software Ballgown [29] can be used for statistical analysis, which has the advantage of processing the output results from upstream software such as StringTie [26].

5 Notes

1. For gram-positive bacteria with stiff cell wall, it should leave out **step 2** and add lysozyme to fracture cell wall. Detailed steps: add 180 μL of buffer (20 mM Tris, pH 8.0; 2 mM Na_2-EDTA; 1.2% Triton; 20 mg/mL of final concentration lysozyme) and incubate at 37 °C for more than 30 min.
2. If RNA needs to be removed, add 4 μL of RNase A (100 mg/mL), shake for 15 s, and wait for 5 min.

3. If the solution is not completely clear, it indicates that cells are not completely lysed, which would affect the quality of extracted genome DNA.
4. If residual ethanol is present, it affects the subsequent experiments such as digestion and PCR amplification.
5. The volume of the elution buffer should be not less than 50 μL and the pH value of water used should be in the range of 7.0 and 8.5.
6. Be as quick as possible at this step after adding TRIzol to prevent the degradation of mRNA.
7. From this step, all operation steps should be RNase-free.
8. The volume of aqueous RNA is about 600 μL corresponding to 1 mL of TRIzol, and the volume of aqueous RNA took could be less than 600 μL in case of some organics under water phase.
9. Add 0.5 mL isopropyl alcohol corresponding to 1 mL of TRIzol.
10. If the concentration is too low (less than 500 ng/μL), the SpeedVac can be used to concentrate it. However, the timing is critical because RNA should not be completely dried.
11. For the super mix, there must have another 3 min for denaturation at the beginning to activate the enzymes.
12. SYBR green is a 50× stock solution. Remember to dilute to 1× in 10× PCR buffer for using at 1×.
13. Newer programs do not have this pause option.

References

1. Harris LM, Desai RP, Welker NE, Papoutsakis ET (2000) Characterization of recombinant strains of the clostridium acetobutylicum butyrate kinase inactivation mutant: need for new phenomenological models for solventogenesis and butanol inhibition? Biotechnol Bioeng 67 (1):1–11
2. Tomas CA, Welker NE, Papoutsakis ET (2003) Overexpression of groESL in clostridium acetobutylicum results in increased solvent production and tolerance, prolonged metabolism, and changes in the cell's transcriptional program. Appl Environ Microbiol 69(8):4951–4965
3. Dragosits M, Mattanovich D (2013) Adaptive laboratory evolution—principles and applications for biotechnology. Microb Cell Factories 12:64
4. Lee H, Popodi E, Tang H, Foster PL (2012) Rate and molecular spectrum of spontaneous mutations in the bacterium *Escherichia coli* as determined by whole-genome sequencing. Proc Natl Acad Sci U S A 109(41): E2774–E2783
5. Yinan W, Xinhui X, Chong Z, Heping L, Liyan W (2017) Recent progress on atmospheric and room temperature plasma (ARTP) biobreeding technology, instrumentation and its industrialization. Biotechnol Business 1:37–45
6. Zhang X, Zhang C, Zhou QQ, Zhang XF, Wang LY, Chang HB, Li HP, Oda Y, Xing XH (2015) Quantitative evaluation of DNA damage and mutation rate by atmospheric and room-temperature plasma (ARTP) and conventional mutagenesis. Appl Microbiol Biotechnol 99(13):5639–5646
7. Ren F, Chen L, Tong Q (2017) Highly improved acarbose production of Actinomyces through the combination of ARTP and penicillin susceptible mutant screening. World J Microbiol Biotechnol 33(1):16
8. Gu C, Wang G, Mai S, Wu P, Wu J, Wang G, Liu H, Zhang J (2017) ARTP mutation and genome shuffling of ABE fermentation symbiotic system for improvement of butanol production. Appl Microbiol Biotechnol 101 (5):2189–2199

9. Cao S, Zhou X, Jin W, Wang F, Tu R, Han S, Chen H, Chen C, Xie GJ, Ma F (2017) Improving of lipid productivity of the oleaginous microalgae *Chlorella pyrenoidosa* via atmospheric and room temperature plasma (ARTP). Bioresour Technol 244 (Pt 2):1400–1406
10. Cheng G, Xu J, Xia X, Guo Y, Xu K, Su C, Zhang W (2016) Breeding L-arginine-producing strains by a novel mutagenesis method: atmospheric and room temperature plasma (ARTP). Prep Biochem Biotechnol 46 (5):509–516
11. Ma Y, Yang H, Chen X, Sun B, Du G, Zhou Z, Song J, Fan Y, Shen W (2015) Significantly improving the yield of recombinant proteins in Bacillus subtilis by a novel powerful mutagenesis tool (ARTP): alkaline alpha-amylase as a case study. Protein Expr Purif 114:82–88
12. Li X, Liu R, Li J, Chang M, Liu Y, Jin Q, Wang X (2015) Enhanced arachidonic acid production from Mortierella alpina combining atmospheric and room temperature plasma (ARTP) and diethyl sulfate treatments. Bioresour Technol 177:134–140
13. Zhang X, Zhang XF, Li HP, Wang LY, Zhang C, Xing XH, Bao CY (2014) Atmospheric and room temperature plasma (ARTP) as a new powerful mutagenesis tool. Appl Microbiol Biotechnol 98(12):5387–5396
14. Qiang W, Ling-ran F, Luo W, Han-guang L, Lin W, Ya Z, Xiao-bin Y (2014) Mutation breeding of lycopene-producing strain *Blakeslea trispora* by a novel atmospheric and room temperature plasma (ARTP). Appl Biochem Biotechnol 174(1):452–460
15. Fang M, Jin L, Zhang C, Tan Y, Jiang P, Ge N, Heping L, Xing X (2013) Rapid mutation of Spirulina platensis by a new mutagenesis system of atmospheric and room temperature plasmas (ARTP) and generation of a mutant library with diverse phenotypes. PLoS One 8(10): e77046
16. Joachimsthal E, Haggett KD, Jang J-H, Rogers PL (1998) A mutant of Zymomonas mobilis ZM4 capable of ethanol production from glucose in the presence of high acetate concentrations. Biotechnol Lett 20(2):137–142
17. Yang S, Land ML, Klingeman DM, Pelletier DA, Lu TY, Martin SL, Guo HB, Smith JC, Brown SD (2010) Paradigm for industrial strain improvement identifies sodium acetate tolerance loci in Zymomonas mobilis and *Saccharomyces cerevisiae*. Proc Natl Acad Sci U S A 107(23):10395–10400
18. Mohagheghi A, Linger JG, Yang S, Smith H, Dowe N, Zhang M, Pienkos PT (2015) Improving a recombinant Zymomonas mobilis strain 8b through continuous adaptation on dilute acid pretreated corn Stover hydrolysate. Biotechnol Biofuels 8:55
19. Mohagheghi A, Linger J, Smith H, Yang S, Dowe N, Pienkos PT (2014) Improving xylose utilization by recombinant Zymomonas mobilis strain 8b through adaptation using 2-deoxyglucose. Biotechnol Biofuels 7(1):19
20. Liu YF, Hsieh CW, Chang YS, Wung BS (2017) Effect of acetic acid on ethanol production by Zymomonas mobilis mutant strains through continuous adaptation. BMC Biotechnol 17(1):63
21. Bolger AM, Lohse M, Usadel B (2014) Trimmomatic: a flexible trimmer for Illumina sequence data. Bioinformatics 30 (15):2114–2120
22. Langdon WB (2015) Performance of genetic programming optimised Bowtie2 on genome comparison and analytic testing (GCAT) benchmarks. BioData Min 8(1):1
23. Li H, Handsaker B, Wysoker A, Fennell T, Ruan J, Homer N, Marth G, Abecasis G, Durbin R (2009) Genome project data processing S: the sequence alignment/map format and SAMtools. Bioinformatics 25(16):2078–2079
24. Luo R, Liu B, Xie Y, Li Z, Huang W, Yuan J, He G, Chen Y, Pan Q, Liu Y et al (2012) SOAPdenovo2: an empirically improved memory-efficient short-read de novo assembler. GigaScience 1(1):18
25. Haas BJ, Papanicolaou A, Yassour M, Grabherr M, Blood PD, Bowden J, Couger MB, Eccles D, Li B, Lieber M et al (2013) De novo transcript sequence reconstruction from RNA-seq using the trinity platform for reference generation and analysis. Nat Protoc 8 (8):1494–1512
26. Trapnell C, Pachter L, Salzberg SL (2009) TopHat: discovering splice junctions with RNA-Seq. Bioinformatics 25(9):1105–1111
27. Kim D, Langmead B, Salzberg SL (2015) HISAT: a fast spliced aligner with low memory requirements. Nat Methods 12(4):357–360
28. Pertea M, Kim D, Pertea GM, Leek JT, Salzberg SL (2016) Transcript-level expression analysis of RNA-seq experiments with HISAT, StringTie and Ballgown. Nat Protoc 11 (9):1650–1667
29. Frazee AC, Pertea G, Jaffe AE, Langmead B, Salzberg SL, Leek JT (2015) Ballgown bridges the gap between transcriptome assembly and expression analysis. Nat Biotechnol 33 (3):243–246

Index

Michael E. Himmel and Yannick J. Bomble (eds.), *Metabolic Pathway Engineering*, Methods in Molecular Biology, vol. 2096, https://doi.org/10.1007/978-1-0716-0195-2, © Springer Science+Business Media, LLC, part of Springer Nature 2020

MIX
Papier aus verantwortungsvollen Quellen
Paper from responsible sources
FSC® C105338

If you have any concerns about our products,
you can contact us on
ProductSafety@springernature.com

In case Publisher is established outside the EU,
the EU authorized representative is:
Springer Nature Customer Service Center GmbH
Europaplatz 3, 69115 Heidelberg, Germany

Printed by Libri Plureos GmbH
in Hamburg, Germany